CHANGE AND STABILITY IN URBAN EUROPE

Change and Stability in Urban Europe

Form, quality and governance

Edited by

HARRI ANDERSSON
GERTRUD JORGENSEN
DOMINIQUE JOYE
WIM OSTENDORF

Routledge
Taylor & Francis Group

LONDON AND NEW YORK

First published 2001 by Ashgate Publishing

Reissued 2018 by Routledge
2 Park Square, Milton Park, Abingdon, Oxon OX14 4RN
711 Third Avenue, New York, NY 10017, USA

Routledge is an imprint of the Taylor & Francis Group, an informa business

Notice:
Product or corporate names may be trademarks or registered trademarks, and are used only for identification and explanation without
intent to infringe.

Publisher's Note
The publisher has gone to great lengths to ensure the quality of this
reprint but points out that some imperfections in the original copies
may be apparent.

Disclaimer
The publisher has made every effort to trace copyright holders and
welcomes correspondence from those they have been unable to
contact.

A Library of Congress record exists under LC control number: 00109568

ISBN 13: 978-1-138-70689-7 (hbk)
ISBN 13: 978-1-315-20157-3 (ebk)

Contents

PART II: URBAN QUALITIES AND URBAN LIFE

PART III: URBAN GOVERNANCE AND PLANNING

List of Contributors

Harri Andersson, Department of Geography, University of Turku, Finland

Ludger Basten, Department of Geography, Ruhr-University Bochum, Germany

Anne Compagnon, Institut de recherche sur l'environnement construit, Ecole polytechnique fédérale de Lausanne, Switzerland

Giuseppe Dematteis, Dipartimento Interateneo Territorio, Politecnico e Università di Torino, Italy

Karl Otto Ellefsen, Institute of Urbanism, Oslo School of Architecture, Norway

Francesca Governa, Dipartimento Interateneo Territorio, Politecnico e Università di Torino, Italy

Gertrud Jorgensen, Department of Urban and Regional Planning, Danish Forest and Landscape Research Institute, Denmark

Dominique Joye, Swiss information and data archive service for the social sciences, Neuchâtel, Switzerland

Jacques Lévy, Université de Reims et Institut d'études politiques de Paris, France

Lienhard Lötscher, Department of Geography, Ruhr-University Bochum, Germany

John Nousiainen, Department of Urban and Regional Planning, Danish Forest and Landscape Research Institute, Denmark

Wim Ostendorf, Amsterdam study centre for the Metropolitan Environment (AME), University of Amsterdam, Netherlands

Anne Skovbro, Danish Forest and Landscape Research Institute, Aalborg University, Denmark

Ann-Cathrine Åquist, Centre for housing and urban research, University of Örebro, Sweden

Preface:

Claude Jacquier, CERAT, Grenoble, France

Epilogue:

Robert A. Beauregard, New School for Social Research, New York, United States of America

Preface

CLAUDE JACQUIER

This is one of three volumes to be published as a result of the action COST A9 *Civitas – Transformation of European Cities and Urban Governance*. This action was launched in 1995, at a time when urban studies had not yet acquired *droit de cité* in the European Union programmes. This initiative was launched by a number of European researchers. It coincided with the emergence of the urban question, which now pervades many European Union programmes.

Here, there is no need to dwell on Europe's obvious urban characteristics. Nor is it necessary to defend the diversity of the European programmes targeted to parts of the city. It suffices to cite two documents published recently by the European Commission. To some extent, these memorandums reflect the centrality of this set of themes. Notably, the EU does not assert any particular competence in this field.

The first document is entitled *Sustainable Urban Development in the European Union: A Framework for Action* and was published in 1999. It stresses four topics which are central to the Civitas cooperation project: the economy, society, the environment, and urban governance. The second document is more pertinent to the work presented in this book. It relates directly to the 5th Framework Programme. That EU initiative explicitly includes the urban dimension in its 'priority action 4' document entitled *City of Tomorrow and the Cultural Heritage*.

The COST framework is a European mechanism to provide scientific and technical assistance for national research programmes. An essential aspect of that mechanism is to facilitate such initiatives. It covers a wide range of fields and includes about thirty European countries today. The COST framework can only exist by virtue of a bottom-up dynamic which calls for initiatives by researchers of the Member States. Its work consists of mechanisms of exchange and cooperation.

The COST framework has supported the emergence of this set of urban themes and urban governance on a European scale. At the same time, the cooperative project has allowed researchers and research institutions working on these topics in various countries to meet in an atmosphere conducive to the development of fruitful research in the future. That future is already blending with the present. The Civitas action has made it possible to build upon the research done as part of the programme on urban development and urban governance. That research will be continued under the 5th Framework Programme for the next three years.

As a result of the internationalization of economies, the increased world-wide exchange of goods, populations and information, and the comparative weakening of nation-states within a supranational complex, European cities now seem to take the lead in the development of the continent. Although their political and institutional importance has yet to be fully recognized, these changes have undeniably altered their status. Where once they were a mere relay station for state powers, they have become a key element in the structuring and organization of the European territory.

The Civitas proposal was to compare the changes occurring in European cities and to analyse how each city reformulates policies and ways of governing in response to these transformations. The aim was to analyse the new forms of urban governance by using three points of view. The first is by taking a spatial approach to the transformation of cities. The second is by identifying the changes in the economic structure of cities. Finally urban governance is understood by analysing social and urban fragmentation. Each of these three analytical perspectives was entrusted to a working group. They were expected to show how these changes called for a new formulation of the policies and a new design for urban government.

The aim was to answer the questions raised at the outset of the Civitas initiative. Let us take up some of those initial questions here.

If European reality is essentially urban, what is the meaning of the urban notion today? Which new changes have occurred in the urban structures during the last 15-20 years – that is after the intensive urbanization of the post-war period? Which new ways of organizing urban space have developed? Which changes have taken place in urban life? What are the new roles of cities in the economic globalization process? How can cities manage both economic competition and social cohesion? Which impacts do these changes have on the social structures? Is there an increasing social and spatial fragmentation of cities? Which effects do these developments have on the management of cities and urban policies? How do these various changes exacerbate the welfare and fiscal crises that affect the government of cities? Which new forms of urban governance have appeared?

To answer these questions, three working groups were established at the beginning of the Civitas initiative. In the course of the action, more than fifty researchers were involved in these working groups. They came from forty universities and institutions and represented various disciplines: geography, sociology, economy, political science, urban planning, and architecture. In many countries, participation was open to young researchers or post-graduates. Fourteen countries were members of the action: Austria, Belgium, Denmark, Finland, France, Germany, Ireland, Italy, The Netherlands, Norway, Slovenia, Spain, Sweden, and Switzerland.

A collection of books brings together the papers produced by the working groups during this period of time.

One volume, *Governing European Cities: Social fragmentation, social exclusion and urban governance*, tackles a major problem which can affect the basis of the European democracies in the future. The mechanisms of social exclusion, social marginalization, social fragmentation or, simply, social differentiation, are particularly evident in our modern European metropolises. These mechanisms are likely to weaken the fundamental principle of "being together" which characterizes European cities. First, the book clarifies some concepts which make it possible to explore the fragmentation of the urban societies and to identify the diverse forms that it can take throughout Europe, today and in the future. Then this book treats some strategies that have been drawn up to ensure the regulation of these phenomena and to come to grips with their consequences in terms of urban governance.

Another one is entitled *Governing Cities on the Move: Functional and management perspectives on transformations of European urban infrastructures*. It explores the transformation of European cities and their way of governing themselves. The transformation is related to the question of urban infrastructure, giving particular attention to transport and communication systems. Using many examples from European cities, this book focuses on the structuring role of these urban infrastructures on the social, economic, and ecological performance of the cities in the long run. It underlines the fact that these infrastructures support strategies, both from a functional point of view (the question of urban accessibility) and from the perspective of managing and governing urban spaces.

This volume is entitled *Change and Stability in Urban Europe: Form, Quality and Governance*. Its objective is to explain the transformations in the spatial organization of the European cities that have taken place over the past two decades. It tries to show in what ways these transformations force us to reconsider the traditional definitions of the city. It reviews the changes, starting from the processes of spatial fragmentation, the complexification of the systems of governance, and the transformation of relations of the inhabitants to public space. The book highlights the diversity of the urban forms in relation to the changes which affect the cities. On that basis, this book tackles the question of the quality of life in new urban spaces. It goes on to discuss the capacity of urban policies to evoke real changes in the city and to regenerate the systems of urban governance.

The books appear at a point in time when European cities, more than ever, are at the heart of the dynamics of the continent. In fact, they may be right in the path of impending storms. We can only hope that this work will be a starting point for new research programmes on the European urban world.

In conclusion, I would like to thank the members of the Technical Committee Social Science of COST, its successive chairmen, Bo Ohngren and Dirk Jaeger, and the successive scientific officers Michel Chapuis, Henrik Graf, Gudrun Maass, Vesa-Mati Lahti for their help during this programme.

Of course, I would also like to thank the following researchers, who were involved in the action A9 Civitas. I hope I have not forgotten any of them.

Hans Thor Andersen (Denmark), Harri Andersson (Finland), Roger Andersson (Sweden), Ann-Cathrine Åquist (Sweden), Antoine Bailly (Switzerland), Michael Bannon (Ireland), Ludger Basten (Germany), Robert Beauregard (USA), Hubert Beguin (Belgium), Maurice Blanc (France), Kerstin Bodstrom (Sweden), Mario Boffi (Italy), Ivar Brevik (Norway), Jack Burgers (Netherlands), Anne Compagnon (Switzerland), Pasquale Coppola (Italy), Barbara Czarniawska (Sweden), Giuseppe Dematteis (Italy), Frans Dieleman (Netherlands), Martin Dijst (Netherlands), Ingemar Elander (Sweden), Karl Otto Ellefsen (Norway), Heinz Fassmann (Austria),

Jürgen Friedrichs (Germany), Alex Fubini (Italy), Oscar W. Gabriel (Germany), Francis Godard (France), Francesca Governa (Italy), Vincenzo Guarrasi (Italy), Peter Güller (Switzerland), Knut Halvorsen (Norway), Brita Hermelin (Sweden), Wolfgang Jagodzinski (Germany), Hubert Jayet (France), Bernhard Joerges (Germany), Gertrud Jorgensen (Denmark), Dominique Joye (Switzerland), Pavlos Kanaroglou (Greece), Hans Kristensen (Denmark), Klaus Kunzmann (Germany), Isabella Lami (Italy), Jolanda Lesnik (Slovenia), Jacques Levy (France), John Logan (USA), Lienhard Lötscher (Germany), Guido Martinotti (Italy), Erkki Mennola (Finland), Markus Meury (Switzerland), Sako Musterd (Netherlands), John Nousiainen (Denmark), Giampaolo Nuvolati (Italy), Willem Ostendorf (Netherlands), Ursula Reeger (Austria), Paolo Riganti (Italy), Walter Schenkel (Switzerland), Harry Schulman (Finland), Daren Scott (United Kingdom), Anne Skovbro (Denmark), Veronique Stein (Switzerland), Jacques-François Thisse (Belgium), Isabelle Thomas (Belgium), Ronald Van Kempen (Netherlands), Ann Verhetsel (Belgium), Jan Vranken (Belgium), Andres Walliser Martinez (Spain), Ari Ylonen (Finland).

1 Introduction

DOMINIQUE JOYE, HARRI ANDERSSON,
GERTRUD JORGENSEN, WIM OSTENDORF

Presentation

A COST book

This is a special type of book. It does indeed resemble one of the collections of various authors' works that are so commonly published today, but this work is no mere compilation. What it is is the product of a group effort involving numerous exchanges during the meetings we held to examine the urban realities in the four corners of Europe. The contributors, from as far away as Cologne and Barcelona, Grenoble and Antwerp, Naples, Lausanne, Copenhagen, and Stockholm, exchanged ideas and criticised one another with a view to fine-tuning their contributions and preparing them for the collective result.

The most important quality of the book is surely its varied perspectives. It is rare indeed to see a work comprising the contributions of Finns, Norwegians, and Danes, as well as French, Swiss, and Italian nationals, who furthermore have compiled a shared vocabulary from their individual traditions and disciplines. It is even rarer to find the urban realities of these different countries aligned side by side where they reflect, beyond their idiosyncrasies, a certain image of a European city.

That said, and with all due respect for the wealth of domains and experience involved in the project, two questions may still come to mind: firstly, why write a book about cities today, and why specifically about stability and change? And secondly, what are its salient points?

Why a book about the city

Why is it that the subject of the city and urban life is the centre of so much attention nowadays? It could be because of all the attention given to the topic by the media, much of which, by the way, is negative: stories of urban violence, images of exclusion and degradation. But the real reason lies elsewhere: cities have always signified high population density and a multiplicity of functions. They have had a crucial role to play in the development of countries and regions. In short, they represent a great centralising force. This fact alone makes cities the privileged witnesses to the problems of our societies, just as they exhibit the noteworthy qualities offered to dwellers and users.

Without wishing to make a hasty generalisation, let us say that European cities differ to a certain point from their American counterparts. The latter are shaped by the existence of the automobile, and their social problems are concentrated around a decrepit centre, whereas European cities rather tend to revive their historic centres and export certain problems to the suburbs. This simple statement, although somewhat of a cliché, does show clearly that the city and its workings cannot be analysed independently of a wide perspective that comprises the social and the spatial, the economic and the ecological, a perspective that includes ways of acting on this system and in particular on its political elements.

But the book is also about stability and change. What has changed in urban systems? A first answer singles out social change, even though this topic is already much discussed in various domains of the social sciences.

What has changed is first and foremost the relationship towards work, which is becoming more evanescent, evolving from the realm of production to that of communication. This automatically induces change in forms of sociability, which may multiply or on the other hand become strained. In short, a multitude of different categories are being grafted onto the social classes that represent the relationship with the world of production. What description can now be applied to a middle class of employees who contribute to the international system? What label can be given to the categories whose relationship to work is transient? How can we speak simultaneously of the continuity of ancient references, built on a tradition of stratification, and of a city with different sets of rules for different people, whose accessibility by different worlds is becoming a cardinal virtue?

What has also changed is the connection to centrality, today defined by one's participation in a world-wide network rather than by a relationship to the hinterland. This transformation is related to the change in the relationship to work. In this context, it is all the more important to promote daily mobility, or at least distances travelled; daily mobility is the key to solving problems caused by increasingly specialised jobs and expanding urban spread, albeit at a certain cost to people and to the environment. Against this backdrop, the scale of proportions has evolved so that the city can no longer be envisioned independently of an international context. At the same time, paying attention to local issues is fundamental in order to comprehend the process of local government and the shaping of those identities for which the living environment is an important reference. In other words, one of the new challenges of urban analysis is specifically to take into account all the various levels. One example is the problem of public places, which are as much spaces for living and interacting for the neighbourhood as they are attractions for promoting the image of the city.

And finally, a change has also taken place in the role of politics: competition is seen as a model, public policies are being divided into sectors, and public action is no longer perceived as monopolised by the state, but rather as potential for the creation of partnerships. This change is perhaps magnified by the dismantling of institutional levels, which throughout Europe has resulted in a weakening of local authority, without the implementation of a replacement level of government. Herein lies the ambiguity of the concept of governance which designates a form of government specialised in problem-solving, but possibly forgetful of the fact that the basic vocation of politics is also to provide a group context and to create conditions of legitimacy. The change is further amplified by an increase in participation, whose legitimacy is justified not by institutionalised channels such as political parties or unions, but rather by the direct presence of grass-roots movements and associations. This in turn raises the issue of representativeness and the rule of access to politics. In brief, the perception of what is public and what is private has been turned completely upside down.

What has remained stable is perhaps the relationship with morphology: types of behaviour taking place within a defined space, and using this space as a resource and in so doing, being somewhat confined by it. What has also not changed is the valuation of places, although the means of doing so are no longer the same. For example, it is no mere chance that the debate surrounding the "new urbanism" has taken on such importance in certain circles today.

A contradictory spectacle then presents itself: that of cities whose uses are multiplying, whose fabric is disintegrating, but which at the same time are seeking centrality and even sometimes finding it in the model of a compact city. The uses referred to are not only geographical, but also social: how can groups seeking to maximise their landmarks live together in an urban space? And what are the means at the disposal of authorities to guide the development of cities?

Form, quality and governance

These are the problems underlying the structure of the book, which we divided into three parts.

First, the issue of form: undertaking any action with respect to the city implies taking stock of the centrifugal and centripetal forces, and also determining once and for all what constitutes and defines a city. Today, a city is not confined to the limits of its walls, but extends over vast urban areas which are built on a scale of entire regions. This question of urban morphology is not restricted to the city limits, but is also a critical part of the internal structuring of cities. The hypothesis that we are defending is that morphology and urban development are not without effect on the behaviour of inhabitants nor on the functioning of a city.

Next, the issue of quality: the primary concern here naturally is the quality of life of city-dwellers, but a hypothetical urban quality which may affect not only daily life, but also the reputation of the city may also be considered.

Finally, these three problems must be managed, and we must study how proposals for urban development, for changing form and for improving quality can be implemented. This is where methods of regulation, irrespective of their types, are of the essence.

These three elements, form, quality, and governance, are obviously involved in a reciprocal causal relationship. It is not just a question of stating, for example, that depending on the morphology of the city, this or that type of governance will be established, although it is clear that politics can shift the form or the quality of urban life. Similarly, issues relating to the quality of life touch upon social questions which will in turn affect urban morphology and the domains involved in authority and regulation. The urban question is thus dynamic and multi-layered, which again implies resorting to the varied perspectives that we have tried to present in this book.

Of course, not all the topics that may concern a city are present in this work. Some may be surprised not to find in it any reference to the new information technologies, for example.[1] Others will be astonished not to see any mention of social exclusion, or more precisely marginalisation. But this work is one in a series; the issues of mobility and infrastructure and of social questions constitute the core of other volumes.

Part I : Urban Form and Processes of Change

This part gives on a different scale, the importance and consequences of the morphological as well as structural changes in urban form. The contributions will explore these changes on different scales, from the supranational scale to the more local. The main point is that the metropolization process changes both the structure of the urban networks as well as relations in the urban area. In the latter case, urban growth typically takes two related physical forms: (1) extending the urban margin through rural land conversion and suburbanization, and (2) adapting, modifying, rebuilding, or replacing existing developed uses and structures. In this case, there are also two levels: one linked to the governance problem of an urban area composed by of a number of communes, the other referring to the quality of urban space inside the cities.

Some basic questions correspond to this thematic. Let us begin with the identification of the urban systems at different scales, as an explanation of the capacity of the authorities to manage the city. At the local level, continued sprawl implies a reflection of what constitutes the centrality and what (and for whom) constitutes the accessibility of such spaces. The consequences of uneven urban development contributes to and reinforces physical isolation, social alienation and segregation, political fragmentation, and spatial distortions in labour markets. Disinvestments in the built environment and redistribution of urban activities to the fringe areas are leading to an impoverishment of urban management and to a reduced capacity of local governments to provide necessary social services and infrastructure.

At the same time, we need to take into account the morphological characteristics, i.e. to know more precisely the influence of architectural form and urban landscape on urban identity, which is the influence of physical form on lifestyles and housing demand, and the role and characteristics of the public space. We may also

---

[1] Without wishing to summarise the debate on the new information technololgies, research shows that although these innovations theoretically lead to decentralisation, they are often simply added onto existing structures, both because of their use of infrastructure and because their use requires human potential that is often located in urban centres.

ask, "Is the revitalisation of obsolete industrial areas, sometimes located near the centre of the cities a means of changing the urban dynamic?" At a more global scale, how can we conceptualise urban dynamics with regard to the new spatial division of labour, flexible production systems and location, and the new role of urban systems (cities as actors in the local/global development processes, keeping in mind the role of urban milieu, social networks, local identity and local development)?

The regulatory environment for urban development in Europe shows tighter restrictions e.g., on urban sprawl (through such practices as shopping center restrictions in new areas, rural land conservation policies, heritage preservation, greenbelts and green zones, and the restrictions on development rights). The aim of these policies is to encourage more intensive reuse of existing built environments, and to produce more "compact" and tightly integrated urban forms. In contrast, the uneven forms implied by the strong urban fringe ("post-suburban") developments and edge-city models, which are the logical outcomes of the prevailing culture of American urban development.

The comparative theme is addressed by **Jacques Lévy**, who analyses "urbanness" after seven principles in different cities. The conceptual discussion between a-priori urbanness and a-posteriori urbanness, and later between relative and absolute urbanness, offers tools for understanding "what is urban". Levy uses the results and conclusions of VillEurope, which since 1993 has in different empirical case studies tried to identify measurement instruments for urbanness and better definition of "Europeanness" of the city. One general hypothesis has been to present two major models of city in the contemporary world: the "Amsterdam model" and the "Johannesburg model". These two models of urbanism represent a major trend in urban organizations in their different indicators of urban structure and activities. The measure of the urban in general or the "urbanness" in particular cannot in fact be distinguished from the perception of the urban reality, and the changes which are most important. In this sense, Levy's chapter is also an orientation for the general outline of the book.

The making of new, multi-centred urban patterns is considered by **Giuseppe Dematteis** and **Francesca Governa** in the context of the characteristics of urban regions in Europe. The authors base their empirical data on the Italian outer urbanization, *città diffusa*, and compare other European situations to this model. The chapter provides an approach to new urbanization in various European countries. The authors state that "the new forms of urban expansion are, in part, the evolution of other urban forms". The model of the *cittá diffusa* is presented as an interesting type, linking elements of centralisation and decentralisation. This is clearly seen in the map presentations that show Italian urban development on different scales. Dematteis and Governa not only point out the physical features of *cittá diffusa*. They also discuss the social, economic and political meaning of multi-centred urban patterns (such as the consequences of these models for quality of life and governance processes, also by referring to the conditions by which a local "milieu" can be constituted).

The conflicting growth areas in urban edges and in inner city areas are at the heart of **Harri Andersson**'s chapter of the development of Finnish cities. Andersson illustrates the dynamics of spacing and timing urban development, where the spatial form of the city is the result of a variety of social, economic and political processes. These processes are "conflicting" in their spatial nature, revealing both centralisation and desetralisation developments in urban areas. The main question in this chapter concerns "dispersal versus internal growth". In Finland, there are many features in the prospects for technological and economic development in the near future which suggest that the dispersal of urban structures will continue. Important questions for urban governance are: "Is urban growth in terms of population or economic activity a necessary prerequisite for dispersal or can growth proceed without dispersal?"; "What part is likely to be played in this trend by the high-rise residential suburbs of the 1960's and 1970's and by other urban fringe areas?"; "And, and to what extent will the developmental forces exert differentiated impacts upon cities in different parts of the country or, more importantly, in different parts of the cities, e.g. their centres and peripheries?"

The last contribution in part I is **Karl Otto Ellefsen**'s chapter discussing alternative models for spatial urban development. Using the city of Oslo and Viken region as case study Ellefsen tests alternative models (policies) to explain existing spatial patterns and transformation trends in urban development. Ellefsen basic question is: "What is new, and what is not new in urban structures?" This theoretical background allows possibilities to formulate development trends, the concept and the role of the city, ideals for urban development, and limitations and potentials of political influence on urban development. The author uses two typologies of urban ideals ("compact city" and "cittá diffusa") to analyze alternative spatial models for future development of the Oslo/Viken region. The models applied yield interesting results to evaluate future urban development particularly concerning urban form and structure as well as urban planning practices.

Part II: Urban Qualities and Urban Life

Whereas urban form can be seen as the "what" of urban development and urban governance as the "how", urban quality and urban life must be understood as the "why". The quality of cities and of urban life is - or ought to be - the ultimate goal for urban governance and should be pursued in the development of urban form.

Urban quality is a very broad construct indeed, embracing aesthetic, cultural, and social issues as well as urban functionality, environment and economy. Quality can only be understood and measured in a human context: single elements of quality - or single qualities - can be measured on the basis of scientific methods, but the discussion and decision about urban quality as a whole must be a political one. However broad the construct, this book deals only with a certain part of urban qualities and urban life, namely those most closely related to urban form and function and to planning process, especially environment, daily life in cities, use of public space, and democracy and participation. These issues are treated in Part II, but they reappear in several of the articles of parts I and III.

One of the most significant debates about urban development and quality in recent years concerns the potential benefits for environment and urban life of compact urban development. In most of the post-war period, relatively low-density development has been generally accepted as the most desirable form of urban development. During the latest 15 years, however, a growing number of researchers have pointed to the fact that compact urban development seems to be environmentally efficient. This position has been advocated from both the theoretical and empirical points of view.

Environmental argumentation has been supplemented by arguments about urban quality, liveliness and economic performance. This position is not new: Jane Jacobs, for example, preached high densities and mixed use development back in 1962. However it was never part of the "politically correct" truth until the European Commission's *Green Paper of the Urban Environment.* This publication re-launched the idea that the European cities should seek to develop virtues of mixed use, medium densities and a well-developed urban life.

An opposing standpoint in this discussion has been that compact urban development would reinstall the shortcomings that urban planning has been combating over the past 70 years (congestion, local environmental stress, etc.) This position also maintains that dense urban environments are not desireable as housing environments and, moreover, that dense development is not necessarily sustainable. **Anne Skovbro's** paper, **The Compact City and Urban**

Quality, assesses the changes occurring from a densification process in an already dense urban area with a case study of an inner
city district in Copenhagen. Skovbro demonstrates that a densification process actually takes place in the district. Skovbro describes
the spatial, functional, environmental and social consequences for
the local environment, discussing the planning system and its ability to cope with such processes.

One central dimension of urban quality - and certainly a most
visible one - is the quality of public space. Especially in dense urban structures, where a minimum of private open space is provided, this measure of urban quality is very important. Public
space can be found in many forms and in many uses: the busy
street for traffic and shopping, the quiet "plaza" where cafés offer a
drink, or the urban park providing the city dweller with a touch of
nature. Empirical research indicates that both parks and urban
public spaces will be greatly utilised if they are provided and of
good quality. Examples show that considerable sentiment can be
attached to certain public spaces, and that plans to change them
can cause alarm or even uproar, in some cases forcing local governments to abandon the plans. In **Dominique Joye** and **Anne
Compagnon's** article: **"Public Places and Urbanness"**, the concept
of public space is discussed, both as a spatial phenomenon and as a
social and political meeting place necessary for the continuous development of democracy and social coherence in cities. The second
part of the paper presents the results of a survey on the actual use
of three public spaces in Geneva. Joye and Compagnon investigate
the use of three different public places in terms of the social characteristics and political preferences of the users and analyse the
role of public places in social integration in the city.

Local and international debates about urban quality are often
based on the debaters' different views, not only upon preconditions, means and results of policies, but also upon the entire project
of urban quality itself. In recent years, the perspective has shifted
from a purely local focus on local environmental issues and functionality of neighbourhoods to a global perspective, including the
entire resource consumption of the city. This is in line with the present political agenda, but neither has the privilege of being "true".
Hence, while it is difficult to scientifically measure individual urban qualities, it is impossible to reach a scientific truth about urban
quality as a whole.

One way to decide about urban quality and future city development is to ask people what they want by ensuring a broad participation in the planning process. **Basten and Lötscher's: "Participation and Quality of Life - Experiences with Local Agenda 21"**

describes the history of public participation in planning - from a mainly reactive protest against single plans in the neighbourhood perceived by the neighbours as a threat to local qualities - to proactive participation of people in decisions about the future of their city in an "Agenda 21" process in Munich. This process has been open and flexible and involved new actors on the scene. Besides the tangible outcomes - consisting mainly of "Agenda 21" projects - the process has given the city a political platform for proactive participation in urban development processes.

Part III: Urban Governance and Planning

As has been touched upon above, the goal of urban planning links the three parts of this book. Urban planning tries to influence, direct or accommodate processes of urbanisation in order to create urban structures that improve the quality of urban life; that is· to promote economic growth and/or a better life for the residents. Although this goal is not always stated explicitly, it forms at least the rationale or hidden agenda of urban planning and of forms of urban governance.

Urban planning is traditionally directed at the improvement of people's quality of life, as is the case with the construction of new housing areas in urban or suburban settings or urban renewal in order to improve the housing situation in deprived neighbourhoods. However, urban planning also tries to reach general goals, such as a reduction in mobility, the creation of sustainable situations, or the preservation of natural areas. One may also think of the creation of infrastructure which promotes economic growth, such as highways or a larger airport. The creation of high quality areas for the location of (foreign) firms and/or foreign investment and the revitalisation of cities and urban areas constitute further examples of the promotion of general rather than individual interests. The same applies to attempts to create a social mix in urban neighbourhoods in order to prevent the growth or continuation of an urban underclass living in ghettos.

With respect to physical planning as well, the problem of scale is urgent. Since the process of suburbanization has created metropolitan regions, the problem has arisen of the difference between the scale of the functioning of urban regions and the scale of cities and municipalities, being the formal authorities for planning and political decisions. This situation has resulted in the concept of "free riders", people or areas functioning within the sphere of a metropolitan region, but failing to assume political responsibility for this fact by giving up political autonomy and become a part of the political body of the metropolitan region instead. The desire to adapt the political level to the level of the daily urban region, of the housing market and of the labour market has created the ongoing discussion on metropolitan governments. In general, one can say that there appears to be no easy solution to the need for metropolitan government.

The contributions in this third part of the book focus on these kinds of problems. Two articles treat the goals of urban planning and the tension between the promotion of individual and general interests. Two others concentrate on the problem of metropolitan government.

Ostendorf discusses changes in urban planning in the Netherlands. Urban development and spatial planning in the Netherlands are highly related to the efforts and effects of state-behaviour in all fields. During the 1970s, state-related planning reached its maximum levels with the development of the "new town policy". The objectives were to prevent overcrowding in the cities and to avoid uncontrolled urban sprawl in suburban areas. Social housing was the most important instrument for achieving the goals: the government invested in the construction of single-family houses, which were strongly desired by the individual households. In terms of realising individual interests, the policy was a success. Nevertheless, it was criticised for not ensuring general interests: since economic activities did not follow the population, the new town policy was considered to have failed. New towns became dormitory cities, and at the same time the donor cities lost too large a share of their populations. Commuting increased substantially, as did traffic congestion and environmental problems.

The compact city policy appeared to be the right answer. However, this new type of policy was developed within a framework of decreasing state-involvement and increasing power given over to market forces. The control opportunities of the state were reduced. The promotion of the interests of individual households and firms is not a clear part of this policy. General interests, such as compact city development and a reduction of mobility, are dominant. Since market forces frequently tend to operate on a demand-led basis, suburbanisation of households and firms is likely to be far more important than the policy allows. The compact city policy therefore also tends to fail.

The contribution of **Åquist** analyses how the function of planning in a Swedish town has changed over time. The Master Plan of 1955 tried to accommodate people according to the everyday life attributed to them. In those days, families were the dominant type of households, and the plan tried to cater for their needs, although a single-family house is not considered to be appropriate. In this last respect, the Master Plan tried to direct the life of people in order to prevent "inappropriate behaviour", such as consumption of too much living space, resulting in too low densities. Here it is not the promotion of a better life for the residents which is the goal, but rather preventing negative effects for society. Over time, the promotion of individual interests - accommodating people based on their personal needs - has become less important. It is replaced by general goals, such as preventing some sorts of behaviour, not because individual people do not want to behave that way (in fact they often want to), but because of factors considered to be negative for society in general, such as high mobility. Here too, the idea of planning as accommodating people's everyday life is adapted to a representation of how everyday life should be organised in order promote a general goal of sustainable development. In sum, it is a change from accommodating plans to preventing plans.

In his contribution **Nousiainen** discusses the role of the government in guiding urbanization. What is the correct administrative unit for dealing with specific tasks: the search for the proper administrative level, one that optimizes effectiveness, efficiency and democracy. Especially problematic is the governance and management of functional urban regions with the help of metropolitan government. While the principle of subsidiarity requires the appropriate level of government, this does not solve the problem, because the question remains how to apply this principle in individual cases. The Greater Copenhagen Area is treated as a case study, showing that there is no easy solution to the problem of metropolitan government.

Like Nousiainen, **Lévy** also focuses on the problem of adapting the administrative level to the spatial scale of the problems. He concentrates on France, considering this case as an exaggerated example of other European countries. Lévy asks why a metropolitan government is at all needed. In his answer, he concentrates on the unity of the daily urban region, asking for a corresponding metropolitan government. He then considers the opposition against such a territorial reorganisation of powers. Institutional inertia is part of the explanation, but allowing individual municipalities the option not to associate in such a government cannot be justified by invoking democratic rights. The state of law, democracy and justice, the principles of the political society, require a metropolitan government. This need is even more urgent in a globalizing world with high-speed mobility and communication.

Part I
Urban Form and
Processes of Change

2 Measuring Urbanness

JACQUES LÉVY

Introduction

Many contemporary works suggest that we are now experiencing a victory of the 'urban' and a defeat of the 'city'. This is an apparently easy way to sum up a current situation, whereby urban areas are ending into something fundamentally new. Doing this, however, we might take the risk of giving up trying to think together realities that, despite of their differences, belong to the same realm, to the same rationale. Should we abandon any hope of constructing common tools for the whole spectrum of urban phenomena, today and in the past, here and elsewhere? Should we downgrade our pretensions to a mere approach of what happens in the city and ignore the simple but upsetting question of what exactly do we mean by the word "city"?

The state of the art in this field shows two main orientations: 1) the measurement of morphological agglomerations, generally based on the continuity of a built-up area; 2) various functional indicators, often referring to daily commuting and sometimes other criteria. In many European statistical surveys we observe different mixtures of both approaches; they are the pressure of a third: the web of administrative districting, municipalities or other local jurisdictions. Everybody acknowledges that neither a purely morphological, nor a simple set of mobility data, even less a primary opposition between the official eponymous 'city' and its 'suburbs' can count for the present-day mutations. But how can we cope with them while maintaining a comprehensive approach to the urban phenomenon?

Seven principles

The following principles comprise an attempt to go beyond this unsatisfactory 'bricolage' and to set up a more consistent way of considering urban spaces.

Adopt a basic and universal definition of urbanness

We must again discover the framework proposed by the most relevant works of the Chicago School, namely Louis Wirth's famous manifesto 'Urbanism as a Way of Life' (1938). This text emerged as a result of three main sources: 1) Georg Simmel approach of urban life as a dimension of a post-communautarian environment; 2) a concern of urban space among German geographers (unlike the French who were still bogged down in their ruralist perspective); and 3) the first empirical sociological space-oriented studies carried out in Chicago and other North-American cities. Wirth delivers a simple but fundamental message: the city is a specific geographical layout, based on co-presence. By the association of density and diversity, a city affords an effective solution to transportation and communication, the other ways to cope with the obstacles that distance raises to social interaction. As a result, density and diversity represent a good measure of urbanness, which can be defined as what makes a city a city.

Distinguish a priori urbanness from a posteriori urbanness and urban capital

Traditionally, cities are classified according to their economic functions or by the distribution of their social groups. A more efficient approach would require a prior study of the potentialities offered by the basic layout of the urban 'form', which is actually much more than a contentless form. The compactness of a built-up area, the level of accessibility between the various places which, together make up a city, comprise relevant criteria to characterise, at a primary stage, an urban area. Thus, two metropolises such as Cairo and Los Angeles, which are virtually incomparable in terms of marketable production or social structure, become relatively similar if we use indicators of spatial pattern such as dwelling density or quality of transportation system. This is the purpose of the approach to a priori urbanness: analysing the geographical assets of a city independently from their social, economic and political use (a posteriori urbanness).

Beyond these two space-oriented measurements, a third set of indicators can be explored. It is called 'urban capital' and consists of a non-spatial overall assessment of the city according to classical economic, sociological, or political efficiency criteria. This approach can be seen as a *translation* of the assets of a specific geographical situation – a *place* – into some universal indicators which can be applied to any kind of social aggregate. Thus, the productivity of a city, be it solely economic or more comprehensive, can be compared to that of other urban or non-urban social layouts. For instance the Gross Urban Product (GUP) per capita of a particular city can be compared to that of the entire country where this city is located (see below 'Two global models...').

Distinguish relative urbanness from absolute urbanness

The larger the city, the greater the mass of possible interactions, according to the formula $I = n (n - 1)/2$, where n is the number of urban 'items' (inhabitants, buildings, organisations,...). Yes, but... this depends upon the nature of the items co-present in an area and on their propensity to interact. In a mining town, for instance, or in any single-activity (manufacturing, tourism, commerce) aggregate, the lack of diversity works as a limiting factor. The effective sum of possible encounters is a decisive element, too: a sprawled city broken up into secluded communities takes less advantage of its size than a smaller one, with a more effective proximity between its different neighbourhoods. Hence we can study separately, at least at a first stage, mass effects (absolute urbanness) and the part of an urban phenomenon which is independent from its size. Measuring separately from realities is a condition for exploring their possible similarities. Many preliminary observations show that although relative and absolute dimensions of urbanness are methodologically independent, urban mass and urban style are correlated. Thus, in a given set of cities, density grows with population.

Distinguish pedestrian metrics from automobile metrics

Research on transportation has often been oriented towards the analysis of multi-means combinations. This can be justified to improve the efficiency of the overall transportation system. If the point is the "intelligence" of the city and the measurement of urbanness, another framework is required. The two main categories of urban movement – 1) by car; 2) by any means, including mass transit, where a pedestrian can remain a pedestrian – have very different effects on urban life. Both *metrics* have a strong power of structuration on distance, proximity, and on the urban space as a whole. Pedestrian metrics allow different kinds of multisensorial interactions and enhance the role of public spaces, where unexpected productive encounters ('serendipity') are more likely to occur. Beyond its direct impact on the destruction of public spaces (transformation of the basic street network into a single-use area, parking spaces, freeways and their buffer zones) and on the ensuing loss of density, many studies have shown the strong link between the use of automobile and the achievement of an urban way of life whose classical hallmark is the single-family owned house. In this model, largely achieved in most North American cities and in European peri-urban areas, the relationship individual/society is the search for family self-reliance which entails a tendency toward a certain level of self-segregation from other parts of the urban society, separation from other social groups and from other urban functions. Exceptions like Oslo, where individual housing did not prevent major use of mass transit systems, are not common, or, at the opposite end, like São Paulo, where mistrust toward urban mix is associated with automobile and high densities, confirm the importance of transportation metrics. Cars and pedestrians make, at the same time, two different cities.

Encompass simultaneously territory and networks

Contemporary cities are multi-speed areas with a 1 to 100 (strolling walk to high speed train) ratio from the slowest to the fastest. This is a first in urban history. Unlike former situations, where this differentiation represented the mere projection of the socio-economic or socio-political structure (fast-dominant *vs.* slow-dominated) and a simple way of rank social strata, today virtually each urban citizen has the opportunity to experience this diversity of metrics. Everyone who lives in a city actually lives in many different cities. This superposition of various metrics presents a challenge to our understanding capacity and an invitation to go beyond the Euclidean space representation. More specifically, a city appears as a mix of two main families of spaces: territories, with continuous and exhaustive distance units; and networks, with discontinuous and lacunary distance units. Distinguishing territories from networks allows us two perceive their dialectics. A street and its sidewalks can be both a network (as a transportation infrastructure, including for walking), and a territory (as a narrow and volatile but efficient crossing zone and meeting place). Cerdà, more than one century ago, boldly expressed it, the quality of a city lies only on the frail balance between *vías* and *entrevías*, that is, between what connects and what is connected. The complexity of the network/territory relationship is well revealed by the two different sorts of territories generated by car and by pedestrian metrics. In the first case, the complete appropriation of all kinds of road networks, including the 'capillary' web, leads to an 'oil-stain' spreading territory with a growing pressure to unify it at the highest speed level, the breaking point being the transformation of all streets into motorways. As for pedestrian metrics, the limit point would be a complete articulation of all operating scales, from walk to train, each of them occupying its comparative advantage niche: unlike the unlimited territorial fusion at a weak level of relative urbanness offered by automobile, pedestrian metrics create an archipelago of strong territorial units. What is true at an intra-urban scale is even more obvious among inter-urban relations. Continuity of built-up area may be an illusion, as it is from New York to Philadelphia or from Paris to Le Havre when it does not match a functionally unified area. Conversely, a powerful network can allow high-intensity exchanges when the connected nodes are strong territories, as it is the case between Tokyo and Osaka or between Paris and various cities of the European Ridge.

Design a point-to-point measurement

Especially in Europe, we are used considering a standard urban pattern with a historical core surrounded by suburbs. In the same spirit, we often emphasize home-to-work commuting and ignore the trips aimed at other purposes. In doing so, we take the risk of neglecting powerful emergent phenomena like new peripheral centres (identified as *edge cities* in North America) or non-job, leisure or commerce-oriented polarities (like shopping centres or "Cineplexes"). To encompass such processes, we must address the geographical layout of a city no more as a ex-ante datum but as a ex-post outcome, at stake in our attempt to measure urbanness. The point is making those facts *visible*, and we may achieve this goal by adopting a spatial background as neutral as possible. The reference map on which we are to mark the measured data must have no preliminary borders of what would be the urban area. We must also cautious regarding the side effects of administrative boundaries (their size and their shape) on the results of our study and, of course, rule out any special treatment of the eponymous municipality. Moreover, we have dire need of a thorough reflection on a cartographic language appropriate for taking account of the intermingled metrics (car and pedestrian, territory and network) of urban spaces.

Build ready-to-use indicators

A measurement tool is meaningless if not usable. The statistical apparatuses of various countries are not always compatible, and even in the European Union much remains to be done to provide an efficient and versatile set of information on urban phenomena. Research programmes require either the use of the available data or 'home-made' productions made from scratch and with limited resources. In both options, a certain level of approximation seems unavoidable. If we are aware of these limiting effects, we can cope with them and try to bridge the gap between concepts and indicators. We can give quantitative data a relevant meaning within our theoretical framework, even if these had not been designed for it. For instance, dwelling density can be used with some qualification to sum up density in general, transportation times to approach accessibility, and land and real estate values to address a complex issue: the way members of an urban society assess the value of different places in a city.

An empirical study on Los Angeles, Paris, Tokyo

Since 1993, VillEurope, a research group funded by the French CNRS, has carried out various programmes to establish and empirically test measures of urbanness, and to try to provide a better definition of 'europeanness' of the city. The most recent stage of this research included a comparative study on three metropolises: Los Angeles, Paris, and Tokyo. The goal was to approach point-to-point 'a-priori urbanness', in its relative and absolute dimensions, using data on the compactness of the urban area and accessibility by car and mass transit.

- Here are some major outcomes of this still unfinished research programme.
- In the three cities surveyed, the best accessibility level from a given point is better in dense areas. The highest ratio between the accessible population within one hour and the population of the whole continuous built-up area is obtained in central Paris (almost 100%), both by car and public transit system. The highest absolute number of accessible population is found in central Tokyo (over 24 million), thanks to public transportation. In spite of its dense freeway grid, Los Angeles hardly reaches 60% and 8 million, in limited areas located in downtown and its surroundings.

- In terms of compactness, the three cities show comparable data, such that Los Angeles attains a surprisingly good score, partly because it is spared by space 'moth-eating', a countryside-city mixture often typical of the peri-urban areas in countries with old rural civilisations such as France and Japan. This is also to be related to Los Angeles' clear-cut boundaries, deriving from its location between ocean and desert. Paris and Tokyo present a more complicated layout, with sinuous townships along valleys. This good performance of Los Angeles is made possible by the fact that density has been approximated with a built-up area (to eliminate the gap between residential density and daytime density). Absolute figures are not as high for the Southern California metropolis.

- Public transportation shows its evident superiority in providing a general point-to-point accessibility in a large city. Density appears as the only good solution to achieving good accessibility in these over-ten-million-people cities. However, density makes adequate car access almost impossible, as we can observe in common several-hour traffic jams in Tokyo. In the intermediate case of Paris, the relatively smooth road flows are a result of the high quality of its mass transit system. When public transportation supply is weak, as in outer ring/outer ring routes, the road network reveals its inability to cope with a growing traffic demand.

- The analysis of relative urbanness shows a steep and apparently paradoxical contrast between car and pedestrian metrics. Because of their cost, mass transit lines make up a lacunary and hierarchised network but they provide a transportation supply which is roughly evenly distributed within the densely urbanised areas. Conversely, in spite of the appearance of equal access to a unified road network, the car accessibility map shows significant differences from one specific spot to another one close by. This situation derives from the fact that there is no simple ratio between the existence of a densely urbanised area and the supply of a massive, fast and fluid road routes. It may even be the opposite: the I-10/I-405 interchange in Los Angeles, the most famous, almost permanent gridlock in the United States, generates an obvious traffic viscosity, without comparison with the viscosity observed in Tokyo Central Station, probably the biggest railroad hub in the world.

- The relationships between this measurement of a priori urbanness and other aspects of urbanness have not yet been completely explored. Let us just point out one issue, related to 'segregation', an aspect of a posteriori urbanness. At the metropolitan scale there is no evidence of a clear relation between accessibility, and more generally a priori urbanness, and the location of poorer populations or 'minority' groups. Watts and South Central in Los Angeles and Paris' northeast suburbs enjoy a very good level of accessibility to the remainder of the urban area. This necessitates a more cautious linkage between social 'exclusion' and spatial confinement.

Two global models of city

The world of cities – as the World full stop – is increasingly differentiated and unified. This process of globalisation, which began with the Great Discoveries, has known many detours and reversals. For the moment, this integration is marked by the growing dominance of hierarchised networks on the traditional jigsaw puzzle of geopolitical states, by the relevance of core/periphery relationships between social groups, organisations and places, and by the emergence of a global civil society still largely deprived from its political counterpart. As those places in the World which are the most connected with each other, cities are deeply involved in this uneven but generalised interaction. We cannot be surprised when we observe strong similarities from one city to another throughout the planet. We can thus notice the global or practically global diffusion of some urban objects, often born in the West, like American-style Central Business Districts or European-style low-rent housing projects, shopping centres or pedestrian streets, urban motorways or underground railways all common landmarks of big cities and, in a reduced form, of smaller ones, too. Generic places have multiplied much faster than during the 'forced globalisation' of European colonisation. Simultaneously globalisation means the emergence of places, that is complex spatial configurations which are not liable to move. When traditionally mobile items (commodities, capital, people) become even more mobile, the contrast with non-mobile, permanent items appears sharper. As actors in the globalisation process, cities play their part with a script which is a unique combination of individuals, groups, skills, organisations, powers, time and space.

If places matter, it is not, or at least not only, because of an alleged 'resistance of places' but because singularity and universality walk at the same pace. In this sense, regionalising of the world of cities may produce strange outcomes. If we adopt the criterion of the relative importance of public mass transit, Sub-Saharan Africa and North America would be on the same side of weak public policies in a context of low-density urban areas, while Europe and East Asia would find themselves on the opposite side.

At this stage, we may hypothesize only two major models of city in the contemporary world. We can call the first one the 'Amsterdam model' and the second the, 'Johannesburg model'.

Table 2.1   Two models of city

	"Amsterdam"	"Johannesburg"
High density	+	–
Compactness	+	–
Good intra-urban accessibility	+	–
Strong pedestrian metrics	+	–
Co-presence housing/jobs	+	–
Diversity in activities	+	–
Social mix	+	–
Strong intra-urban polarities	+	–
High Gross Urban Product per capita	+	–
Positive self valuation of all urban area	+	–
Self identification of the urban society	+	–
Urban-scale governance	+	–

The basic idea of this dichotomic typology is that all components of relative urbanness (independently from urban mass) are 1) positively correlated with each other; 2) largely embodied in actual cities. Of course, Amsterdam is not the perfect expression of the Amsterdam model, nor Johannesburg the achieved incarnation of the Johannesburg one. But each model represents a major trend in urban organisations. In the Amsterdam model, the advantage of concentration achieves its highest level, i.e. co-presence and interaction between a maximum of social operators. Only the individual is entitled, with his/her housing, to a right of remoteness (privacy). In the Johannesburg model, separation in any form structures an urban space made up of a mosaic of homogeneous and partly independent neighbourhoods. In spite of its specific character – the ghastly radicality of apartheid and its legacy – the actual city of Johannesburg is representative of many other cities, including many in Southern and Western North America. With its strong identity, its bourgeoisie conveying through centuries a design whereby free-market and solidarity, social hierarchy and citizenship were deeply interwoven, its constant and consistent urban policy, Amsterdam (like Barcelona, Hamburg or Bologna) provides a sharp and perhaps excessively favourable image of what is the most specific of the European city.

Both models can be seen as the expression of major civilisational options and, at this level of generality, ranking them would be highly questionable. In the 1997 ranking given by the development level composite indicator (Human Development Index, HDI) proposed by the UNDP (United Nations Development Programme), Canada and France come in first and second respectively. Now Canadian urban spaces pertain more or less to the Johannesburg category, while French cities, with qualifications, are rather akin to the Amsterdam 'template'. This apparent confusion can be interpreted in two ways. Either urbanness has a weak impact on the overall performance of a society; or – this would be my own hypothesis – when an urban society experiences a poor situation regarding its specifically *urban* assets, other social assets can to a certain extent make up for it.

Table 2.2   The 25 richest cities, ranked by GUP, and their OPR

	City (morphological agglomerations)	Population (thousands)	Gross Urban Product (GUP) (G$)	Over Productivity Ratio (OPR)
1	Tokyo	29 317	1443.8	1.246
2	New York	24 310	829.2	1.249
3	Osaka	15 011	628.7	1.060
4	Los Angeles	14 539	457.4	1.152
5	Paris	9 513	361.4	1.515
6	Nagoya	6 851	291.3	1.076
7	Chicago	8 991	273.6	1.114
8	San Francisco	5 567	213.9	1.407
9	Washington/Baltimore	5 706	212.1	1.361
10	London	8 017	208.9	1.394
11	Seoul	18 942	193.6	1.090
12	Boston	5 346	173.5	1.188
13	Hong Kong/ Shenzhen	7 101	140.0	(a)
14	Miami	4 385	132.8	1.109
15	Essen	4 669	128.9	1.000
16	Dallas	4 031	124.4	1.130
17	Detroit/Windsor	4 097	120.2	(a)
18	Buenos Aires	11 757	111.5	1.180
19	Toronto	4 587	107.4	1.181
20	Milan	3 850	102.1	1.385
21	Hamburg	2 151	101.5	1.709
22	Taipei	7 583	100.2	1.100
23	Houston	3 381	98.5	1.067
24	São Paulo	16 333	94.6	1.563
25	Mexico	17 738	93.5	1.588

(a): international agglomeration. Source: Geopolis/VillEurope, 1998.

This latter interpretation seems confirmed by the measurement of Gross Urban Products carried out by François Moriconi-Ébrard (Geopolis database) for VillEurope. The ratio GUP per capita/GDP per capita of the respective country, called OverProductivity Ratio (OPR), shows a clear advantage to 'Amsterdam'-type cities, mostly in European metropolises, while some cities in North America (New York, San Francisco), in Japan, and Latin America, too, as do Los Angeles, Miami, and Dallas, obtain more disappointing scores.

These data suggest that the overall efficiency of 'Amsterdam'-type cities is better than that of the 'Johannesburg'-type ones, including in a specifically economic dimension. Strangely enough, many comparative works (see for instance, Sassen, 1991) yet exclusively polarised by the financial component of growth and development and despising too 'literary' concepts like knowledge or creativity, just seem to neglect this basic but hardly disputable monetary accounting.

Taking urban space seriously, we could not be surprised by a better urban performance of cities that attain a better level of urbanness. Because of their European style urbanness, European cities remain the most productive spatial machines in the world. Conversely the Johannesburg model typically encompasses a kind of city where urbanness is everyday and everywhere challenged by its opposite. No wonder, in those conditions, that, in comparison to other spatial options, the urban choice cannot gain its full momentum.

References

CATTAN Nadine et al. (1992) *Le concept statistique de la ville en Europe*, Luxembourg: Eurostat.

CERDA Ildefonso (1867), *Teoría general de la urbanización y aplicación de sus principios y doctrinas a la reforma y ensanche de Barcelona*, 3 vol., Madrid.

SASSEN Saskia (1991) *The Global City*, New York: Princeton University Press.

VILLEUROPE (1995) *Urbanité et européanité*, research report for CNRS, Paris.

VILLEUROPE (1998) *Metroparis*, research report for CNRS and RATP, Paris.

WIRTH Louis (1938) "'Urbanism as a way of life", *American Journal of Sociology*, Vol. 44.

Definition of urban areas

A priori, everyone has an idea of what a city is! But in fact, it is not so easy to agree on an operational definition, even for the statistical offices. In Iceland for example, we speak of a city when a municipality has more than 200 inhabitants, in France the limit is hundred times greater, 20000 inhabitants! Such a difference is not arbitrary. It reflects a different meaning of the urban centrality according to national contexts.

Urban realities no more correspond to the simple idea of the city. Nowadays urban areas extend beyond the municipal boundary of the city center. The definition of such an urban area implies a larger perimeter. Most often, the definition of urban areas involves a group of municipalities and does not break up such units. That introduces a first approximation in the definition. But the question remains: how to combine them? Two sets of criteria can be considered:

An approach looking at the built environment, expecting no break between areas. In this perspective, an agglomeration has to be concrete, physically linked and different from the rural areas. This approach was used, with some variations, by the national census offices in Europe from the beginning of the twentieth century up to the present.

With the development of communications and transport, beginning with streetcars, more functional criteria were adopted, showing the links between the different municipalities constituting the agglomeration. The commuting flows were mainly used as a criteria. This approach has been dominant in recent decades.

Even if some efforts toward international standardization have been made, the urban definitions maintained a national character. The different size of municipalities in Europe is also a source of difficulties in aggregating administrative units to form an agglomeration following similar criteria. And the construction of comparable indicators is often difficult: even the definition of an employed worker could vary according to the rules of the statistical office! Now, at the time of European construction, some comparative works have nevertheless been written, in the scientific as well as in the administrative area. In the latter case, the tendency is toward more intensive use of GIS, to begin with satellite pictures to define the most densely built-up areas. This use of modern and sophisticated systems of observation could mean, paradoxically, that a renewed importance is given to morphological criteria, as an international agreement on functional links seems more difficult to attain.

Dominique Joye

3 Urban Form and Governance: The New Multi-centred Urban Patterns

GIUSEPPE DEMATTEIS AND FRANCESCA GOVERNA

Introduction

The more or less "concentrated dispersion" of settlements is the most prominent feature of the recent changes in spatial forms of urban systems. The urban expansion can be considered, in part, as a mere quantitative extension of the previous suburbanisation process, with a wider decentralisation of dwellings and related services and infrastructures. Nevertheless in many ways it seems to differ from the simple urban sprawl or from the extension and merger of contiguous urban fringes. The new urban expansion reveals qualitative changes such as spatial fragmentation, less regular centre-periphery gradients, the weakening of spatial hierarchies in outer suburban areas and the rise of new centrality patterns (fig. 3.1).

Traditional terms like "suburbs", "rurban", "sprawl" and "rural-urban fringe" are no longer appropriate to describe the new forms of outer urbanization. Even classical concepts such as "city region", "daily urban system" or "metropolitan area" now seem unable to fit the complex form of the new urban systems. R. Fishman (1987) describes recent changes in spatial forms of cities by tracing the significance of traditional suburbs back to their origins in the late eighteenth century and then by showing the evolution to the present. Bauer and Roux (1976), J. Garreau, (1991), F. Ascher, (1995), A. Corboz, (1998) and F. J. Monclús, (1998) speak about *périurbanisation, edge city, métapolis, hypercity* or *ciudad dispersa* to summarise the characters of the new nodal and flexible forms of urbanization as well as to stress location choices of inhabitants and the modification of behaviour and social practices in today's larger settlement pattern.

The recent changes in urban systems are made clear in different studies. They are tackled both from a political and operational point of view, in particular in many official EU documents such as *Europe 2000+* (EC, DG XVI, 1995) and the European Spatial Development Perspective, ESDP, (EC, DG XVI, 1998), and from a theoretical and methodological one. The most immediate and relevant problem for empirical research and statistical surveys is the identification, delimitation and measurement of territorial urban units. This problem is now emerging at the EU level as a crucial and has stimulated the need to set up statistical indicators and comparative research projects (Pumain et al., 1992). At the theoretical level, the interpretation of current changes in the urban systems is based on the advanced application of self-organization models and of fractal and catastrophe theory (Wilson, 1981; Pumain, Sanders and Saint-Julien, 1989; Frankhauser, 1994; Allen, 1998).

This paper approaches these issues at a level intermediate between the empirical-statistical and the formal-theoretical approach. It considers the adequacy of the current conceptual categories to deal with the complexity of current outer urbanization in Europe. The empirical analysis is based mainly on the Italian outer urbanization, *città diffusa*, as a model to which other European situations can be compared. In this view, the new characters of urban regions in Europe are presented not only from the physical point of view, but also in terms of social, economic and political features. This approach allows us to explore spatial transformations of urban form as evidence that reveals processes of urban changes at different geographical scales and to define the process of creation of a new multicentred urban pattern. From another point of view, this problem is also tackled by Karl Otto Ellefsen in this book. In this case, general models for urban spatial development – "compact city" and "città diffusa" – are related to alternative policies and different possibilities for action.

The dispersion of urban growth in external areas is not only transforming many urban characters, for example the compactness or the traditional concentric patterns, but also the way in which cities are governed and managed in accordance with the subsidiarity principle and the concept of governance. The implications of previous analysis for urban governance are mainly related to the explicit search for decentralized and flexible planning models and for a bottom-up approach based on local systems as levers of urban policies.

Main characteristics of European outer urbanization

The massive decentralization of population and employment, with its suburbanization of housing, production and service activities, determines the spread of urban regions. With significant variations in different spatial contexts, this process shows the transformation of the urban form of affluent countries during recent decades. In general terms, the new patterns of urbanization in Europe consist of a continuous settlement grid, organized around a large number of specialized nodal foci in a vast multicentred region. The ideal model of these patterns can be defined as a city without centre or as an urban region organised around the scattered fragments of an exploded centre.

The process of formation of this pattern can be divided into two general models. In the first model, urban expansion is seen mainly as a process of urban growth. In that case, the model of outer urbanization takes the form of ring configuration around cities. In the second model, the dispersion of settlement patterns is related mainly to the links between small and medium-sized cities and defines a polycentric reticular model able to extend urban features to a wide area. These two general models can be applied to describe the formal characters of European outer urbanization.

In France, the new pattern of dispersed settlement, called *péri-urbanisation*, was studied in the French cities from the 1970s from the perspective of land use and density around cities of different size, from the largest agglomerations to several medium and small–sized towns (Bauer and Roux, 1976; Berger and Rouzier, 1977; Berger et al., 1980). During the 1970s and the 1980s, *périur-banisation* was described in terms of spatial expansion of urban systems stemming from the external location of residential functions with the rise of the *ville eparpillée*, and considered as the result of a large scale redistribution of population, production and service activities, with a loosening (*deserrement*) of the settlement mesh. In this view, *périurbanisation* marks the transition from an urban polarized space (with a strong hierarchical dependence of the peripheral rings on the city centre) to an integrated space in which the different functions follow different forms in spatial organization, partially freed from the close city centre control.

Figure 3.1 Conceptual categories and forms of urbanization

Urban systems (typologies)			Patterns of city systems	Patterns of decentralisation
Nuclear			Town/country	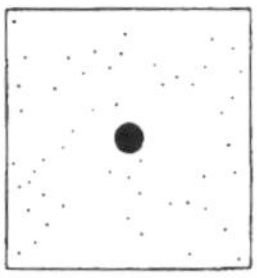 /
Extended	Hierarchical core-periphery	Conurbation	Continuous built-up areas with regular core-periphery gradients	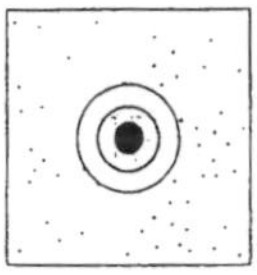 Zonal suburbs
		City region	Hierarchical multinuclear functional areas	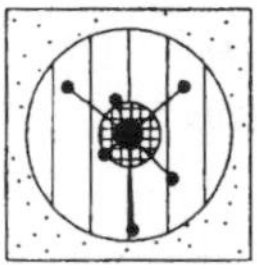 Urban sprawl and multinuclear dispersed city
		Urban Field (J. Friedmann)	Non-hierarchical polycentric integrated areas	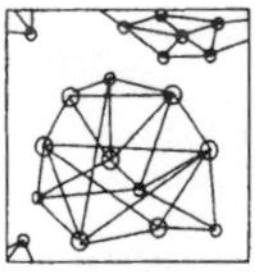 Regional multicentred systems of specialized cluster
	Networked (explosed city)	*Città diffusa Périurbanisation Edge city*	Multi-centred urban patterns	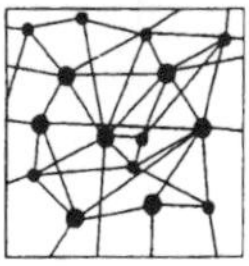 Regional network of single local systems interacting with global networks

Where the density of the pre-existent urban grid is lower, we can find individual urban systems with their extended periurban rings. In some regions, in the Midi or along the Rhone Valley for example, it is difficult to define individual urban systems. In these areas, spatial processes are governed by a more close-knit grid of small and medium-sized cities, specific morphological characters and by development dynamics and the development of mobility infrastructure. Hence, urbanisation thus takes reticular forms in the coastal regions or corridors of intense urbanisation along transportation axes.

Some scholars, particularly Berger et al. (1980), consider *péri-urbanisation* as a form of urbanization dissociated from the traditional form of the city and have pointed out its innovative features in connection with urban expansion. These features derive from the sites in which it occurs (the third urban belt, the urban fringes between city and countryside), and from the social and cultural conditions (in particular, changes in life-styles or in the social composition of inhabitants). In other studies (Dubois-Taine and Chalas, 1997), *périurbanisation* is described as a major change not only in spatial forms of urban systems, but also in the very conception of what is urban; it entails the break-up of traditional hierarchical centre-periphery relationships and the rise of a new pattern of urban centralities.

In Switzerland, we find similar features. Since the 1970s, most Swiss cities and agglomerations are characterised by the demographic reduction of urban cores and by demographic growth in suburban communities. These demographic changes were produced within a context of structural economic changes towards a selective process of tertiarisation and the widespread industrialisation of small towns (Racine, 1992). The Swiss urban system is now "marked by intense flows between the principal urban cores following a network structure which has developed along two principal axes that are conditioned by the relative locations of the two metropolitan regions (Zurich and Lake Geneva basin) as well as the urban region of Ticino, the cross-border metropolitan region of Basel and the federal capital of Berne" (Cunha and Racine, 1996, p. 4). The new urban dynamic in the metropolitanisation process of the Lake Geneva basin thus defines a polycentric model similar to the Randstad in the Netherlands (Leresche et al., 1995). If the Randstad, as an example of a long-term planned region, is a very particular case, other European examples of wide outer urbanization can be found in western and southern Germany, in the British South-East and Midlands and in Italy.

The main features of the new forms of outer urbanization in different European countries exemplify the evolution from a common urban expansion to the new reticular model. This differs from traditional urban sprawl from two main points of view. The first point of view concerns the process of outer urbanization, where the main cause is normally found in the structural changes in the spatial division of labour, influencing the residential choices of the active population. The second point of view concerns the spatial structure. The current European examples exhibit a continuous

merging pattern, marked by a wide range of decentralization processes, the uneven density, alternating high density clusters or corridors and low density rural spaces. It is characterised by spatial
organization around specialized nodal focuses, weakening of core
polarization and rise of a less hierarchical multi-centred network of
social and economic links inside the urban regions.

The Italian città diffusa

New forms of outer urbanization: the Italian debate

In Italy, the recent urban sprawl has been given various names
according to different analytical points of view. Recognition of the
discontinuous, dispersed, low density character of outer urbanization is the common starting point of all studies. However, there are
many differences between them. From the point of view of land
use and density, F. Boscacci and R. Camagni (1994) deal with the
relationships between urban form and sustainable development
and with urban-rural competition for land. From the point of view
of social and demographic changes, urban expansion is considered
not only as a quantitative growth of urbanized areas, but also as
the extension of urban lifestyles to a wide area (Indovina, 1990;
1999). From a physical-morphological point of view, attention is
focused on the various elements of which these forms of urbanization are effectively composed or on the physical characters of settlement patterns and landscape describing the modes of occupation
of the land, open spaces and building typologies as indicators of
the spatial organization of urban expansion (Secchi, 1993). From
the functional point of view, recent outer urbanization is described
in terms of urban regional networks, as systems of relationships
and flows among specialised centres in which reticular intercon-
nected patterns (*reticoli urbani equipotenziali*) take the role of a functional grid of new urbanization (Dematteis and Emanuel, 1992;
Camagni and Salone, 1993).

Taking into account these different approaches, some recent
studies use the term *città diffusa* to summarise different forms and
features of Italian outer urbanization. For example, one basic aim of
the Itaten research project (Clementi, Dematteis, Palermo, 1996)[1]
was to reduce the distance between different approaches, consider-

---

[1]   Itaten - *Indagine sulle trasformazioni del territorio nazionale* - was a research
       project carried out by several Italian universities. The aim was to make a
       survey of the changes to the Italian territory as a resource for the activities of
       the Ministry of Public Works.

ing the interaction between social and physical morphologies. In this case, the interpretation of recent changes in the spatial forms of urban systems recognised not only the extension of suburban growth, but a qualitative change of the economic, physical and social processes of recent urbanization.

Formal attributes, origin and evolution of città diffusa areas

The main formal attributes of the Italian *città diffusa* are the increasing extension of urbanized areas outside and around metropolitan areas (ring configurations) or in open spaces among regional networks of medium-sized cities (areal and belt configurations) or, finally, along major linear geographic features such as coasts, valleys, traffic corridors, etc. (ribbon configurations).

These patterns of urbanization are discontinuous; they include the growth of minor urban systems in rural areas and, at the same time, the suburban location of residential and production functions, the extension of infrastructure networks related to mobility by public and private transport systems. Consequently, the density of building and population is lower than in traditional inner suburbs, with significant densification only around pre-existing nuclear settlement (hamlets, villages, small towns).

The formation of *città diffusa* areas begins in the 1970s from a phase of counterurbanization, i.e. the wide spatial de-concentration and dispersion of demographic growth after the phase of polarized growth in the 1950s and 1960s. This transition from high urban concentration to diffuse growth was related to a change in spatial organization of economic activities with a new form of spatial division of labour and emerging forms of flexible production (Dematteis and Petsimeris, 1989).

In this phase, we can identify two movements: (1) a progressive wide decentralization of the medium level functions heretofore located in big cities and metropolitan areas and (2) a highly selective centralization of innovative and higher level functions. The simultaneous action of decentralization and centralization forces produced a reticular spatial structure, characterised by a redistribution of less specialised functions among minor urban nodes and by the increasing specialization of a few major nodes in higher level functions. A clear example of this process is the formation of functional networks in the west-central Po region with both hierarchical and non-hierarchical links among centres (Dematteis and Emanuel, 1992; Camagni and Salone, 1993) (fig. 3.2).

Figure 3.2 Functional Networks in the west-central Po region with both hierarchical and non-hierarchical links among centres

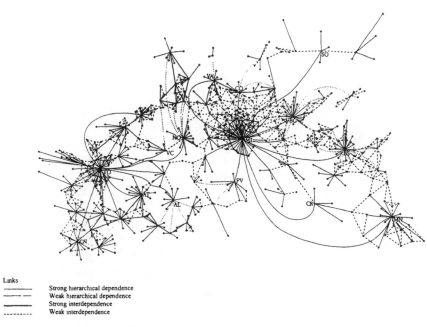

Links
——————— Strong hierarchical dependence
— — — Weak hierarchical dependence
————— Strong interdependence
----------- Weak interdependence

(Source: Dematteis, 1995, p. 100)

During the 1980s, the range of spatial de-concentration processes became shorter, consolidating the above-mentioned forms of "dispersed concentration": rings, belts and ribbons, where a non-hierarchical distribution of functions among the nodes of regional urban networks became normal.

The formation of the *città diffusa* shows that the city, as a no longer simple nuclear or areal entity, exemplifies the evolution from the extension of settlement patterns to a complex multi-centred model combining physical grid expansion and the spatial redistribution of specialized functions.

Figure 3.3 Italy: map of urbanization

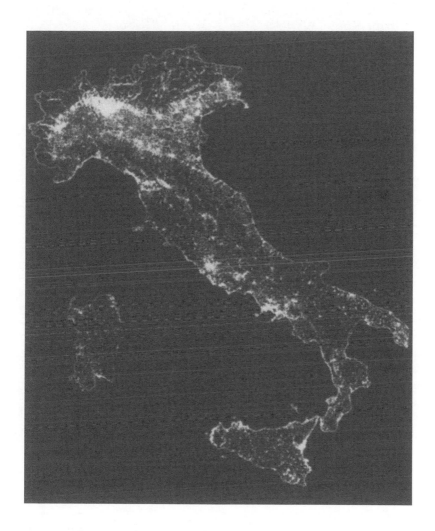

(Source: our processing of Clementi, Dematteis and Palermo, 1996)

The Italian *città diffusa* comprises a variety of geographical situations (Clementi, Dematteis and Palermo 1996; Bonavero, Dematteis and Sforzi, 1999). As shown in the map of urbanization to 1991 (fig. 3.3), its spatial form is evident in many Italian regions,

with different spatial patterns that can be summarized in three
main classes:
- wide rings around major metropolitan conurbations (Milan,
  Turin, Rome, Naples) with prevailing exogenous components, resulting from decentralization of activities and
  population overspill;
- belt and areal patterns forming regional multinuclear systems, with prevailing endogenous components, like Mar-
  shallian industrial districts (e.g., belts along the foot of the
  Alps and Apennines, especially in the Po valley, grid diffusion in Veneto or in the lower Arno valley between Florence
  and the Tyrrhenian coast);
- coastal and corridor ribbons, where physical and infrastructural conditions have an attractive role (examples: corridors
  along major Alpine valleys, ribbons with associated "comb"
  patterns in the middle Adriatic region or following the
  Florence-Rome-Naples line along the internal Apennine ba-
  sins.

Traditional and innovative characters of the new pattern

Some formal attributes and spatial features of the new forms of
settlement patterns are already present in traditional suburban
growth and apparently pursuing traditional urbanization. For example, the physical expansion of the urbanized and commuting
areas or the related extension of infrastructural, commercial and
service networks are already present in traditional suburban
growth. Nevertheless, other features differ from simple urban
sprawl or from the extension and merge of contiguous rural-urban
fringes. They imply innovative changes in previous models of cen-
tre/periphery dependence, the weakness of traditional hierarchies
and the rise of new centrality patterns (Dematteis and Governa,
1999).

One of the most innovative features of *città diffusa* areas is the
integration and complementarity, at the regional scale, among different components of the urban systems: old centres, compact suburbs, outer suburbs, new nodal centres. These components comprise a functional regional-urban continuum, englobing pre-
existent autonomous and self-organizing local systems, particularly
in Italy's many industrial districts. Infrastructural networks and
nodal high range activities (shopping centres, airports, universities,
technology parks, transport nodes, business service centres, etc.)
also play a leading role in structuring the outer urban space.

These areas are marked by a general weakening, with frequent local inversions, of negative centre-periphery gradients concerning land values, population densities, centrality levels, social and environmental conditions. The grids of settlements expand and merge, gradually bounding and eroding rural spaces and reducing farming activities. At the end, the only wide open spaces that remain will be the parks and protected areas, as integrated urban components.

Functional spaces are fragmented and overlapped: no comprehensive self-contained areas emerge from the spatial analysis of flows and other interaction links between centres. Most of the activities are randomly distributed among the nodes of the regional urban network. Each functional unit may have spatial relations of a different range independently of its location. Normally, each sub-centre (node of the regional settlement network) gravitates towards different centres for different functions. So, local nodes can have direct access to global networks (financial, commercial, informational, cultural, etc.), and to autonomous linkages with remote cities and regions, bypassing, when necessary, the closest metropolitan centres. As a consequence, contiguous local nodes can have different trajectories and speeds of growth (or decline) and the urban regional space tends to be fragmented and unevenly developed locally.

From the social point of view, the *città diffusa* is normally less polarized than a large compact city, also because it presents a strong component of self-employed workers and small businesses. Life styles, from many points of view, are similar to the American suburban model: cottages or detached houses, high car mobility, shopping in peripheral hypermarkets and shopping malls, neighbourhood social life and control, etc. Community cohesion and identity are normally present at the village, neighbourhood and small town level, while at higher territorial levels (district, urban region) the urban systems are socially fragmented.

The new forms of outer urbanization show important quantitative and qualitative differences from the traditional suburbs: they have no clear boundaries, they include rural, urban and suburban elements, centres seem scattered over the space, making a multi-nodal urban region that tends to replace the traditional hierarchical patterns.

Multi-centred patterns, local systems and global interactions

In the transition from a hierarchical and polarized structure to a networked and multi-centred urban space, the pre-existent territorial order seems to be fragmenting; at the same time, we can identify some signs of change defining a new organizational form, although not completely clear and univocal. For the present, we do not have analytical models or conceptual images to deal with the spatial complexity of current urban structures. To appreciate this, we must understand the processes that have led to the present forms of urban expansion. They are, in part, the evolution of other urban forms; at the same time, they are marked by some innovative processes.

This view stresses the relationships between different parts of urban systems: recent outer urbanization is not considered as a distinct social and territorial entity, even in contrast to the compact (traditional) city.

The contrast between compact city and *città diffusa* leads to comparison between the outer and the central cities as two abstract models without links to real processes. As we have seen, different components of the larger urban systems, comprising old centres, compact suburbs, outer suburbs and new centres, tend towards a new model of urban spatial organization: a continuous settlement grid organized around a large number of specialized nodal focuses. In this view, the spatial organization seems to be a mesh of different spatial structures *à géometrie variable*, shaped by flows, connecting individual territorial units at different scales, both by proximity links and by long distance ones. In this mesh, each component of urban systems can play a part as an element of a new order in the apparently sprawling disorder, offering relatively stable local "anchorages" to global interactions (Veltz, 1996). However, the incidental presence of these local anchorages does not assure, by itself, any particular social cohesion or operational closure to the whole regional system, but may instead increase its fragmentation.

The new spatial organization reflects social and economic changes, related to the post-Fordist transition. Globalisation and the shift to the information economy and society render special value to large cities as centres for efficient face-to-face exchange and, above all, for the interconnection between different types of global networks (economic, cultural, physical, etc.). Big cities and metropolises become a locus of overlapping webs of relations on different spatial scales. They are the locations of innovative and higher level functions, major airports and high-speed train stations. At the same time, they are also affected by social, economic and environmental diseconomies, for example high rents, congestion, pollution, social exclusion, etc.

These contradictions have some important implications. In particular, certain activities (back office functions, intermediate activities, etc.) tend to be located outside the centre: in corridors leading to the airports, around suburban train stations or in small and medium-sized towns in the surrounding ring. Some medium - sized cities also act as centres of high-technology production and attract longer-distance decentralisation (Hall, 1998) as well as giving rise to specific local development processes. This produces vast multicentred networked regions that no longer have a simple hierarchical dependence on the nearest metropolitan centre, but have relationships with functional networks at local, regional, national and international levels.

Such trends encourage a fragmented view of the urban structure, referring to different relations operating on a variety of spatial scales. At the *global level*, we observe the crisis of industrial activities as historically located in urban centres, the new spatial organization of economic activities and the shift to the information economy (international division of labour, flexible production and location etc.), demographic and social changes and the development in transport and information technologies. At the *local level*, we observe urban concentration and deconcentration processes related to historical, cultural, social, economic, physical and environmental local factors, infrastructures and network accessibility, diseconomies of compact cities, new life styles and other socio-cultural changes.

The complexity of these processes calls for approaches that account for both the pressures exerted by global restructuring and for local specificity. In particular, we need an approach that it enables us to analyze local/global interplay in terms of local specific "responses" to global changes (Amin and Thrift, 1994). These are related to the local capacity to act as a complex and self-organized system and to the new role of urban systems as actors in competing and co-operating networks (Bagnasco and Le Galès, 1997). Hence, the problem is to understand if and when a local space can act as a "place", i.e. as a pro-active local system able to define voluntary actions and strategies.

In this approach, we have to consider local urban systems both from an internal and an external point of view (Conti, Dematteis and Emanuel, 1995). From the internal point of view, local systems are made up of networks of local actors, operating as organizational components of the systems, and of a specific local milieu as the historical heritage of physical and cultural conditions that can provide constraints and opportunities for local development strategies. Public and private local actors, acting only at local level or both at local and global levels, are "close" together if and when they are able to self-organize collective actions to manage local development strategies through the valorization of local milieu conditions. Local systems are thus not only any old aggregation of actors. However, they exist as autonomous entities only if local networks (existing or potential) can self-organize themselves in order to give local specific responses to global changes and to interact with global networks (Dematteis, 1994). In this case, the *local urban milieu* can be seen as the territorial identity of local actors, an identity operating both as a sense of belonging and self-organizing capacity. The milieu conditions provide the endogenous potentials for development and the competitive advantages of each place.

From the external point of view, each local system can be seen as a node of regional networks in terms of spatial proximity interactions and as a node of global networks in terms of long distance interactions. If we combine internal and external points of view, local/global interplay can be seen as the interaction between local nodes and supra-local (global) networks, if and when nodes are pro-active local systems.

Governance issues

A multi-centred urban pattern seems to be replacing the hierarchical organization of centres. Social and spatial fragmentation is increasing. The new spatial organization can be understood as a set of relations with different social, cultural and spatial dimensions, which co-exist in the confined arena of urban areas. Changes in the social, economic and spatial structure of urban systems are reflected in, and reinforced by, considerable changes in the way cities are governed and organised, in accordance with the subsidiarity principle and the concept of governance (Healey et al., 1995).

Urban governance must to deal with three processes that lead to the explicit search for an interactive view of urban policies and for a bottom-up approach to planning. First, it must acknowledge the expansion of the sphere of local political action as involving not merely the local authority but also private and semi-public actors. Second, it must deal with the central role that territorial identity plays both for the "construction" of the process and for the mobilization of public and private actors. Finally, urban governance must confront the current fragmentation in society and space that can be seen in terms of the local nodes/supra-local network interaction (Le Galès, 1998).

Urban governance has to build up a network of different local systems and to arrange a co-operative organization of them. Knowledge of local/global interaction and of the self-organizing principles of different local systems is therefore a major lever for governance of the new forms of urban expansion. However, the self-organizing capacity of local systems is not a property to be taken for granted. If local urban systems are in some way self-organized systems, they can act as connecting nodes of different networks and play an important role in spatial structuring. In that case, the urban network structure develops through self-organization of its parts and components. On the contrary, if urban nodes are not able to give their own "responses" to global stimuli (or to link themselves to other nodes in order to do this), the new urbanisation will appear as a simple quantitative expansion of the traditional suburbs in an enlarged urban region, still dependent on the central city. Thus, we must accept that it is not possible to apply the traditional community model, based on organic territorial social cohesion, to the complex and extended forms of today's urbanization.

The idea of the place-based community has a long tradition in planning thought. However, the present challenge is to find forms of collaborative planning and to reinterpret community to mean "the assertion of the concerns of accomplishing life strategies and *everyday life* in the context of the forums and arenas in which *political community* finds expression, and in which collective activities are organised" (Healey, 1997, p. 126). The new urban communities, at different levels, from the neighbourhood to the region, are thus increasingly dynamic realities to be designed and constructed. The artificial and voluntary nature of today's urban communities does not mean a loss of interest in the geography of cities. If we cannot find already existing territorial urban systems, we can identify the potential ones. There is a geography of milieu conditions supporting social networks, relations among actors, links between public institutions, private organizations and other co-operative spatial interactions, which revolves around common projects of development and territorial transformation revealing the potential for local and regional self-organizing processes.

Different areas of the Italian città diffusa

The central Veneto area between Venice, Padua and Verona is a prototypical example of città diffusa areas. In effect, the term was introduced for the first time by F. Indovina (1990) to describe this context. In that case, urban dispersion depends more on a progressive endogenous densification along road grids of scattered pre-existent settlements than on a metropolitan process of decentralization. However, the process of densification is not homogenous. It is mainly related to some pre-existing features (for example, the settlement diffusion along the layout of the Roman centuriatio). The progressive connection of nuclear settlements determines a discontinuous urban habitat with the extension of infrastructural, commercial and service networks and of urban characters over a wider territory (Indovina, 1990; 1999).

In this area, individual rationality guides most location decisions (Secchi, 1996). The physical and spatial forms of dispersion are also related to specific social and economical forms of local development, such as the industrial districts and SME areas. Thus, the spatial organization in the Venetian città diffusa derived from the interaction between historical and geo-morphological conditions, socio-economic features of industrialization and settlement patterns, dispersion and extensive use of the territory, change in individual and social practices, lifestyles, building typologyes and the central role of private car mobility.

The situation of the città diffusa around Milan is quite different (Boeri, Lanzani and Marini, 1993; Lanzani, 1996). In this area, the dispersion of settlement patterns arises both from traditional suburbanization processes and from the reinforced spatial and functional connections among the small and medium-sized centres of the pre-existing regional urban grid. While the demographic decline of the resident population continues in the conurbation, at the same time the number of city users and metropolitan businessmen is increasing. This demonstrates the leading economic role of the central city and particularly its place in the metropolitan networks of European cities. The città diffusa around Milan is not only related to the growth of a big city but also to the local endogenous development of diversified urban, economic and social structures in the regional space. It has a multiplicity of spatial organizational patterns, with the increasing strengthening of historical nuclear settlements and industrial districts and the rise of a new linear urban continuum along infrastructural and service networks.

In the middle Adriatic region, the settlement pattern of dispersion occurs between the foot of the Apennines and the coast, also supported by the development of mobility infrastructure (Pavia, 1996; Ricci, 1996). Urban dispersion arises both from the transfer or the increase of the inhabitants in small and medium-size centres and from the merger between major and minor towns, linked also to the development of traditional industrial districts. While the historical settlement patterns, with some exceptions was homogeneously distributed in the hills and in the coastal plain, recent expansion develops in linear strips along the coast and valley axes, assuming the form of an urban continuum.

Giuseppe Dematteis and Francesca Governa

References

ALLEN, P. (1998) Cities as Self-organizing Complex Systems, in: BERTUGLIA, C. S., BIANCHI, G. and MELA, A. (Eds.) *The City and Its Sciences*, pp. 95-144, Heidelberg: Physica-Verlag.

AMIN, A. and THRIFT, N. (Eds.) (1994) *Globalization, Institution and Regional Development in Europe*. Oxford: Oxford University Press.

ASCHER, F. (1995) *Métapolis ou l'avenir des villes*. Paris: Odile Jacob.

BAGNASCO, A. and LE GALÈS, P. (Eds.) (1997) *Villes en Europe*. Paris: La Découverte.

BAUER, G. and ROUX, J.M. (1976) *La rurbanisation ou la ville éparpillée*. Paris: Seuil.

BERGER, A. and ROUZIER, J. (1977) *Ville et campagne. La fin d'un dualisme.* Paris: Economica.

BERGER, M., FRUIT, J.P., PLET, F. and ROBIC, M.C. (1980) Rurbanisation et analyse des éspaces ruraux péri-urbain, *L'Espaces Géographique*, 9, pp. 303-313.

BOERI, S., LANZANI, A. and MARINI, E. (1993) *Il territorio che cambia. Ambienti, paesaggi e immagini della regione milanese*. Milan: Abitare Segesta.

BONAVERO, P., DEMATTEIS, G. and SFORZI, F. (Eds.) (1999) *The Italian Urban System. Towards European Integration*. Aldershot: Ashgate.

BOSCACCI, F. and CAMAGNI, R. (Eds.) (1994) *Tra città e campagna. Periurbanizzazione e politiche territoriali*. Bologna: Il Mulino.

CAMAGNI, R. and SALONE, C. (1993) Network urban structure in Northern Italy: elements for a theoretical framework. *Urban Studies*, 30, pp. 1053-1064.

CLEMENTI, A., DEMATTEIS, G. and PALERMO, P.C. (Eds.) (1996) *Le forme del territorio italiano*. Roma-Bari: Laterza.

CONTI, S., DEMATTEIS, G. and EMANUEL, C. (1995) The development of areal and networks systems, in: DEMATTEIS, G. and GUARRASI, V. (Eds.) *Urban network*. pp. 45-68. Bologna: Pàtron.

CORBOZ, A. (1998) *Ordine sparso*. Milan: F. Angeli.

CUNHA, A. and RACINE, J.B. (1996) Towns and metropolitan areas in Switzerland: federalism torn between radical mutations in the work force and conservative patterns of behaviour. Paper presented at the *IGU Commission on Urban Development and Urban Life*. The Hague: 5th to 10th August.

DEMATTEIS, G. (1994) Global networks, local cities, *Flux*, 15, pp. 17-23.

DEMATTEIS, G. and EMANUEL, C. (1992) La diffusione urbana: interpretazioni e valutazioni, in: DEMATTEIS, G. (Ed.) *Il fenomeno urbano in Italia: interpretazioni, prospettive, politiche*, pp. 91-103. Milan: F. Angeli.

DEMATTEIS, G. and GOVERNA, F. (1999) From urban field to continuous settlements networks. European examples, in: AGUILAR, A.G. and ESCAMILLA, I. (Eds) *Problems of Cities: Social Inequalities, Environmental Risks and Urban Governance*, pp. 543-556. México: UNAM, Istituto de Geografia.

DEMATTEIS, G. and PETSIMERIS, P. (1989) Italy: counterurbanization as a transitional phase in settlement reorganisation, in: CHAMPION, A. G. *Counterurbanization*, pp. 187-206. London: E. Arnold.

DUBOIS-TAINE, G. and CHALAS, Y. (Eds.) (1997) *La ville émergente*. Paris: Editions de l'Aube.

EC (EUROPEAN COMMISSION) DG XVI (1995) *Europe 2000+. Cooperation for European Territorial Development*. Brussels-Luxembourg: Office for the Official Publications of the European Communities.

EC (EUROPEAN COMMISSION) DG XVI (1998) *European Spatial Development Perspective*. Brussels-Luxembourg: Office for the Official Publications of the European Communities.

FISHMAN, R. (1987) *Bourgeois Utopias: the Rise and Fall of Suburbia*. New York: Basic Books.

FRANKHAUSER, P. (1994) *La fractalité des structures urbaines*. Paris: Anthropos.

GARREAU, J. (1991) *Edge City. Life on the New Frontier*. New York: Doubleday.

HALL, P. (1998) Cities of Europe. Motors in Global Economic Competition, Paper presented at the ESPD Seminar. Lille: 22-23 June.

HEALEY, P. (1997) *Collaborative Planning. Shaping Places in Fragmented Societies.* London: MacMillan.

HEALEY, P., CAMERON, S., DAVOUDI, S., GRAHAM, S. and MADANI-POUR, A. (Eds.) (1995) *Managing Cities. The New Urban Context*. Chichester: Wiley.

INDOVINA, F. (1999) La città diffusa. Cos'è e come si governa, in: INDOVINA, F. (Ed.) *Territorio. Innovazione. Economia. Pianificazione. Politiche.*, pp. 47-59. Venezia: Daest.

INDOVINA, F. (Ed.) (1990) *La città diffusa*. Venezia: Daest.

LANZANI, A. (Ed.) (1996) Lombardia, in: CLEMENTI, A., DEMATTEIS, G. and PALERMO, P.C. (Eds.) *Le forme del territorio italiano. II Ambienti insediativi e contesti locali*, pp. 73-102. Roma-Bari: Laterza.

LE GALES, P. (1998) La nuova political economy delle città e delle regioni, *Stato e mercato*, 52, pp. 53-91.

LERESCHE, J.P., JOYE, D. and BASSAND, M. (Eds.) (1995) *Métropolisations: interdépendences mondiales et implications lémaniques*. Genève: Georg Editeur.

MONCLUS, F. J. (Ed.) (1998) *La ciudad dispersa. Suburbanización y nuevas periferias*, Barcelona: Centre de Cultura Contemporània de Barcelona.

PAVIA, R. (1996) Marche, in: CLEMENTI, A., DEMATTEIS, G. and PALERMO, P.C. (Eds.) *Le forme del territorio italiano. II Ambienti insediativi e contesti locali*, pp. 193-214. Roma-Bari: Laterza.

PUMAIN, D., SAINT-JULIEN, TH., CATTAN, N. and ROZENBLAT, C. (1992) *Le concept statistique de la ville en Europe*. Luxembourg: Eurostat.

PUMAIN, D., SANDERS, L. and SAINT-JULIEN TH. (1989) *Villes et auto-organisation*. Paris: Economica.

RACINE, J.B. (1992) Urban development and urban life in Switzerland: towards new kinds of socio-spatial logics and polarisations?, Paper presented at the *IGU Commission on Urban System and Urban Development*. Detroit: 3th to 8th August.

RICCI, M. (1996) Abruzzo, in: CLEMENTI, A., DEMATTEIS, G. and PALERMO, P.C. (Eds.) *Le forme del territorio italiano. II Ambienti insediativi e contesti locali*, pp. 215-235. Roma-Bari: Laterza.

SECCHI, B. (1993) Le trasformazioni dell'habitat urbano, *Casabella*, 600, pp. 44-46.

SECCHI, B. (Ed.) (1996) Veneto, in: CLEMENTI, A., DEMATTEIS, G. and PALERMO, P.C. (Eds.) *Le forme del territorio italiano. II Ambienti insediativi e contesti locali*, pp. 128-147. Roma-Bari: Laterza.

VELTZ, P. (1996) *Mondialisation. Villes et territoires. L'économie d'archipel*. Paris: P.U.F.

WILSON, A. G. (1981) *Catastrophe Theory and Bifurcation Application to Urban and Regional Systems*. London: Croom Helm.

4 New Spaces of Urban Transformation: Conflicting 'Growth Areas' in the Development of Finnish Cities

HARRI ANDERSSON

"We are today at the beginning of truly new economic era. It has transformed both what the people do and the places where they live, in ways which a mere twenty years ago would have seemed unimaginable… thanks to these economic changes, place has changed its meaning".
(Richard Sennet "Something in the City")

The context of change

Urban development can be driven by a series of interrelated processes of change: economic, demographic, political, cultural, technological and social (fig. 4.1). These processes also result in some important changes in the character and dynamics of the urban system (the set of towns and cities within a region or nation according to hierarchical systems or networking systems); within cities these processes generate changes in patterns of urban land use (e.g. patterns of transportation and communication), in the built environment (e.g., creation of edge cities and theme parks as new urban

45

realities, and at the same time diminishing the value of old urban areas and traditions), in the social life and social ecology (influencing the social and demographic composition of neighbourhoods), and in the nature of urbanism (the forms of social interaction and ways of life that develop in urban settings). Some of these outcomes may be perceived as problems by certain groups within society. Government policies, legal changes, city planning and urban management may eventually address those problems, often resulting in changes that in turn affect the dynamics that drive the global trends and the overall urbanization processes (cf. Knox 1994, p. 9).

To understand major trends characterizing urban development, it is essential to recall the scale and pace of the restructuring of the international economy. During the 1950s and 1960s, international trade, capital investment, and labour migration patterns contributed to rapid economic growth in the Western industrial nations. This occurred alongside the shift away from industrial production towards services, particularly sophisticated business and financial services, as the basis for profitability. As this shift began to transform occupational structures, the decline in manufacturing jobs, combined with the increased globalization of the economy, contributed to a destabilization of the relationship between business, labour, and government. In Western societies, *deindustrialization* led to a widespread, systematic disinvestment in the nation's basic productive capacity, and to devastating social impact on the local communities when plants shut down or moved their operations elsewhere (see Bluestone and Harrison 1982, pp. 6-7).

During the process of deindustrialization, it became increasingly difficult to extract profits from mass production and mass consumption. This led many enterprises to seek profitability through serving specialized market niches ('just do it'-policy). Instead of standardization in production (cf. 'Fordism'), profitability demanded variability and *flexible production systems*. The net result has been labelled by scholars *disorganized capitalism*, not so much because of the lack of organization or purpose in business, government, or labour, but because of the contrast with the orderly interdependence of all three during the organized capitalism phase (see e.g., Lash and Urry, 1987). The significance of this evolution of megatrends for urban development is fundamental. Each new phase of capitalism saw changes in what was produced, how it was produced, and where it was produced. *These changes called for new kind of cities, while existing cities had to be modified.* At the same time, of course, cities themselves played important roles in the transformation of capitalist enterprise (cf. Knox, 1994, p. 10).

Figure 4.1 Urbanization: processes and outcomes

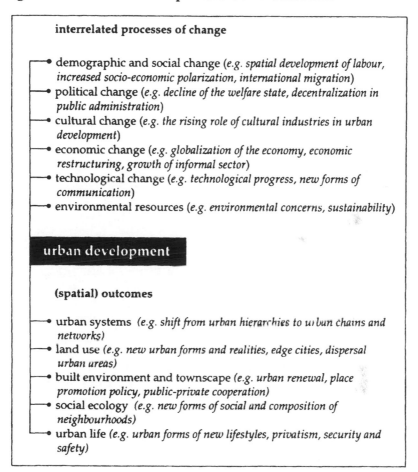

interrelated processes of change

- demographic and social change (*e.g. spatial development of labour, increased socio-economic polarization, international migration*)
- political change (*e.g. decline of the welfare state, decentralization in public administration*)
- cultural change (*e.g. the rising role of cultural industries in urban development*)
- economic change (*e.g. globalization of the economy, economic restructuring, growth of informal sector*)
- technological change (*e.g. technological progress, new forms of communication*)
- environmental resources (*e.g. environmental concerns, sustainability*)

urban development

(spatial) outcomes

- urban systems (*e.g. shift from urban hierarchies to urban chains and networks*)
- land use (*e.g. new urban forms and realities, edge cities, dispersal urban areas*)
- built environment and townscape (*e.g. urban renewal, place promotion policy, public-private cooperation*)
- social ecology (*e.g. new forms of social and composition of neighbourhoods*)
- urban life (*e.g. urban forms of new lifestyles, privatism, security and safety*)

Economic restructuring has two meanings: one as itself, economic restructuring, and the other, more narrowly, as industrial restructuring. The first connotes broad changes in the economy, extending beyond the sphere of production into distribution, finance, governmental relations, and the labour process. The second refers to transformations in the relative importance of industries, what has come to be termed the manufacturing-service shift (see Beauregard 1989, p. 7). At the local level, this shift has in some cases led to environmental crises and speculation (land purchases and demolition of existing built-up urban areas), to fiscal crises (a stagnant local economy becomes a part of strong industrial growth, which means the growth of public infrastructure provision and public expenditures) to financial crises ('hidden roots' of deindustrialization) and to strong private sector growth (flexible accumulation).

Deindustrialization has greatly changed cities and the lives of their inhabitants. It is a process which is a combination of conflicting processes, but it is nevertheless only one of the factors changing the city. As mentioned above at the heart of the dynamics that drive and shape the development of individual cities lie economic changes. The sequence and rhythm of economic change will be a recurring theme as we trace and retrace the imprint of urbanization and urban change. According to Castells (1992, p. 5) the formation of the global economy is one of the major structural trends of our epoch. It is an economy where capital flows, raw materials, labour markets, commodity markets, information, management, and organization are internationalized and fully interdependent throughout the planet.

Tied to arguments concerning spatial deconcentration are hypotheses about *the international division of labour*. Two major points are important in evaluating the role of cities in the changing international division of labour (see Glickman 1987, pp. 66-67). First, there has been a vast transformation of the world economy in that energy price increases, new patterns of investment and migration, and the interpenetration of markets have had profound and uneven effects on national economies. There have also been employment reductions in traditional manufacturing and shifts to the service and high-technology sectors. Second, the third industrial revolution (involving electronics, biotechnology and information processing), it has encouraged deurbanization and the dispersion of urban populations. This is because technology and the maturation of product lines have promoted both standardized work and dispersed job sites, while changing business strategies and organization have allowed firms to seek less urbanized locations in urban fringes.

Figure 2 outlines the phases of Finnish urban development under organized and disorganized capitalism. These phases are also typical for other countries, only the scale and timing differ. In major Finnish cities, the population and industries grew especially during the time of the 'Great Migration' between 1950 and 1970. Employment was concentrated in a few large enterprises, whose production began to slow down in the 1970s. In many cities the textile industries began to decline in the 1970s, the purge continued in the 1980s; the metal industry and the paper industry changed their production strategies.

Figure 4.2 The stages of development of Finnish cities during the 1900s

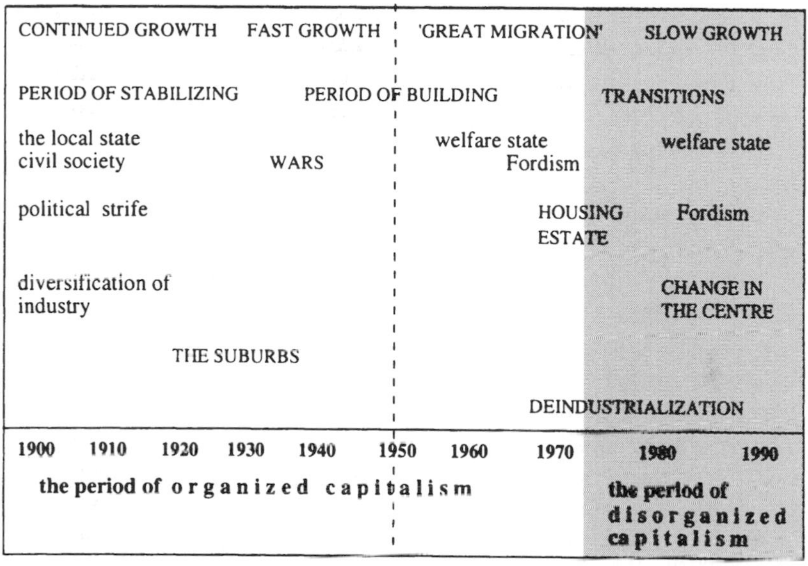

CONTINUED GROWTH	FAST GROWTH	'GREAT MIGRATION'	SLOW GROWTH
PERIOD OF STABILIZING	PERIOD OF BUILDING		TRANSITIONS
the local state civil society	WARS	welfare state Fordism	welfare state
political strife			HOUSING ESTATE — Fordism
diversification of industry			CHANGE IN THE CENTRE
	THE SUBURBS		
			DEINDUSTRIALIZATION
1900 1910 1920 1930 1940 1950 1960 1970 1980 1990			
the period of organized capitalism			the period of disorganized capitalism

Many cities could be described as a modern/industrial city, which in the early 1980s began to go through another transitional period characterized by strong *deindustrialization*. The process of deindustrialization took place at the time when the Finnish welfare state was still being enlarged and improved. The following stages can be distinguished in the process of deindustrialization in Finland: STAGE I (mid-1950s to mid-1970s), which describes relative deindustrialization, STAGE II (mid-1970s to mid-1980s), which describes absolute deindustrialization in all industries, and STAGE III (mid-1980s onwards), which describes more radical absolute deindustrialization (reduction of the work force in all major industries, severe reduction of the work force of all large factories, many of which close down).

Urban outcomes

The spatial form of a city is the outcome of a variety of social, economic and political processes, and the factors contributing to it reflect specific historical circumstances which are part of a larger social reality but which can explain the changes which have taken place in the urban landscape at particular moments in time. Any city will contain examples of physical and social environments which have been created at different periods in time and under different historical conditions. The economic and social processes affecting the structure and functioning of the society in which the town is located determine the nature and extent of the spatial variants to be found within its economic and social functions, and they influence the urban landscapes inside the city. The power structures in society and the allocation of resources are the keys to understanding and interpreting the processes of change in the urban landscape. In his numerous works published between 1975 and 1980, Pahl combines the view of various groups exercising different degrees of power within society with the concept of urban managerialism. Pahl's 'managerialist thesis', derived from urban conflict theory and notions of the post-industrial state, provides a framework for research which points to the existence of a set of factors governing the allocation of urban functions (comprising the representatives of the building companies, property agents and local authorities). By studying the activities of these people and the institutions they stand for, one can move towards an understanding of the functioning of urban markets.

Urban models conceived within the managerial frame of reference take account of the historical dimension to a certain extent and relate to other, broader social processes. Nevertheless, doubts have arisen regarding the capacity of such an approach to comprehend the real nature of society and its influence upon urban processes and the socio-spatial forms within cities. The development of a political economy approach has meant the adoption of a more critical view of changes in urban structure. Instead of institutional conflicts and constraints, urban growth should be viewed as one aspect of a more extensive process of social development, and an urban structure as a product of this development at a given point in time. This also means that 'the faces of power' are more complicated than they are assumed to be in the case of institutional explanations. The spatial restructuring of the city is organized by various 'place entrepreneurs' who practice the politics of local economic development by forming growth coalitions. In their book *Urban Fortunes: The Political Economy of Place*, Logan and Molotch (1987) speak about 'systemic power', which is a result of business people's continuous interaction with public officials. The organization of the growth coalitions ('the growth machine') included anybody who became an entrepreneur in a particular place: politicians, local media, public utilities, financial institutions, even including universities and cultural organizations.

Over the last decade many urban researchers have detected a profound change in the nature of Western societies and global economic relations. Much of this debate revolves around three related issues: the relationship between time and space, the potential of politics, and the construction of identity (Keith and Pile 1993, p.1). There are a many different kinds of approaches using time-space relations as a background for urban studies: Torsten Hägerstrands time-geography (showing time-space routines as a part of every-day life), Anthony Giddens (1979) concepts of time-space distancia-tion (where he tries to connect presence and absence or more gen-erally local and global), David Harvey's (1985) time-space com-pression, capital flows and urbanization, and finally, Allan Pred's (1991) Stockholm studies showing the entrance of modernity in one of the capital cities of Europe using the time-space framework. The second issue, the potential of politics has been discussed in the con-text of regenerating urban areas, and especially of waterfront areas (the regeneration of London Dockland has become a classical ex-ample how an urban renewal process has produced numerous studies of changing urban politics and urban planning). The third issue, the construction of identity, is closely related to the potential of politics connecting urban politics to new ideas of the city (cf. place promotion).

As pointed out in table 1 in most cases the initiators of com-mercial restructuring in Finnish city centres have been financial institutions and building companies, which means obvious links with capital and speculation. As an ideology, inner city renewal produces a better commercial climate and healthy residential areas in places which had previously been degenerating, and profit in places where there had previously been poverty. However the ide-ology tells us nothing of the dynamics of capital (or flexible) accu-mulation which lie behind the process. The illusion of urban re-newal as a form of integration in which the basic attribute of the urban space is a 'richness and variety' of relationships between people from different social and consumption groups collapses because of solutions dictated by the power of capital (property and land owners).

These changes have revealed different kind of 'spatial prac-tices' based both on production of urban space and consumption of urban space. Production-side trends include the increased vertical integration among development companies, the housing industry, real estate companies, architectural practice, etc. They also include increased involvement of conglomerate corporations and financial institutions, public-private cooperation, and increased product differentiation and niche marketing. On the other hand, consump-tion-side trends have meant increased socio-economic polarization, increased significance of commodity aesthetics, new class fractions, identification and reinforcement of consumption communities, and social (re)construction of place. The consumption-side trends also mean a new kind of urban life . The consequences of production-side trends have created new spaces which are 'socio-cultural' flows of urban transformation concerning technological changes and emergency of high-tech industries as well as a new information society, international development of cities, new forms of urban culture, new consumption patterns and shopping districts, and areas of ethnic minorities.

Figure 4.3   Conflicting 'growth areas' (socio-cultural flows) in the inner city of Turku

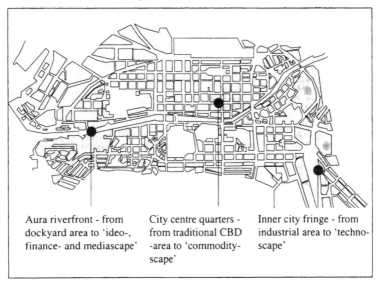

Aura riverfront - from
dockyard area to 'ideo-,
finance- and mediascape'

City centre quarters -
from traditional CBD
-area to 'commodity-
scape'

Inner city fringe - from
industrial area to 'techno-
scape'

Particular criticism has been levelled at the means used for the commercial renewal of inner city cultures. This 'exclusive' type of development pushed the mechanism of democratic participation to one side in favour of private investments and a monolithic 'middle class invasion' (cf. Holcomb & Beauregard 1981, p. 67). In many Finnish cities, the urban planning required for restructuring ('reurbanization policy') has erred on the side of allowing the private sector too much scope for deciding on the organization of urban space. The result is that the new 'megastructures' generated by this exclusive process represent the interest of investors, builders' agents, estate agents, financiers and big business, ending up with a range of services aimed at only one type of consumer, elite commercial environments, over-specialized shopping centres which offer their services on a selective basis only to high-income consumers. These shopping centres, referred to as 'attractions', but they serve for most people as objects of sightseeing trips (cf. 'captive markets') rather than as functional parts of the city.

Table 4.1 New shopping malls in the Finnish city centres

city	city centre mall (name and floor area in square meters)	opening year	initiator(s)
Helsinki	Forum, 20 000	1985	bank
	Kluuvi, 23 000	1989	bank
Turku	Hansa, 65 000	1985-88	property owners
Varkaus	Forum, 5 000	1986	building firm
Lahti	Forum, 7 000	1987	building firm
	Citypalatsi, 6 000	1989	building firm
	Hansa, 8 000	1992	building firm
Tampere	Koskikeskus, 23 500	1988	building firm
	Tullintori, 25 000	1991	central firm
Kuopio	Savon City, 7 500	1988	enterprisers
Lapeenranta	Iso-Kristiina, 15 000	1988	building firm
	Kauppakanava,15 000	1990-94	property owners
Jyväskylä	Torikeskus, 20 500	1988	bank
	Are-kortteli, 8 000	1992	bank
Kotka	Kotka cbd, 12 900	1988	bank
Pori	Be-Pop, 8 000	1989	bank
	Iso-Karhu, 8 000	1991	speculators
Rovaniemi	Sampokeskus, 12 500	1989	insurance company
Vaasa	Rewell-center, 33 600	1990	central firm
Kokkola	SM-keskus, 13 500	1991	property owner
Lohja	Lohjantähti, 6 000	1991	city of Lohja
Seinäjoki	Torikeskus, 19 200	1992	building firm

At the level of economic activity, the problems focus upon the rapid rise in urban land prices and the uncertainty attached to business management under the new conditions now prevailing. Commercial functions aimed at traditional forms of consumption cannot compete with the shopping precincts in the city centres, for urban land prices are highest in precisely those places where extensive changes in retail markets are taking place. One example of this is the 'Forum' mall in downtown Helsinki, construction of which raised rents in that block by over 200 % in real terms between 1980 and 1986, almost twice the increase experienced elsewhere in the city centre. On the other hand, the construction of new inner city shopping areas well away from existing city centres leads to economic and spatial dislocation. The principles of inner city renewal have included the preservation of the old structure where at all possible, which implies that the construction of large, uninterrupted shopping areas in old city centres will require a sufficiently large centre in order to achieve both an intensive concentration of commercial functions and preservation of the existing forms. It has even occurred in some cases that commercial megastructures have had the opposite effect of that intended, i.e. they have not revitalized the service sector economy of the city centre on a broad scale but have milked its remaining commercial strength, giving rise to an imbalance in the internal hierarchy of retail trading in the city as a whole (claims of this kind have been made regarding both the Forum area in Helsinki and the Hansa centre in Turku).

Most of the old industrial buildings or districts in the inner city area of Turku are owned by the city council itself or by real estate investors, as summarized in Table 4.1. Only a few premises are owned by industrialists. Typically, the industrialists have either sold their industrial spaces or established commercial companies for the purpose of negotiating good zoning contracts for future redevelopment or recycling of these areas. The private sector has also been active in taking over industrial premises predominantly occupying or located adjacent to sites which offer the greatest private sector development potential and financial returns - in most cases in the city centre or waterfront area (cf. Loftman, 1994, p. 1). The power relations that lie beyond these reproduced urban landscapes and the meaning of 'urban function' also reflect future forms of urban living – different from the typically gentrified urban areas (see Andersson, 1995).

Spacing and timing in urban development

Two different concepts influence on urban restructuring: spacing ('scales' of urban restructuring, figure 4.4) and timing ('biases' of urban development, figure 4.5). As a result of the post-war macroeconomic change, Finnish cities grew and expanded very rapidly. The scope of planning control concentrated on the physical and technical aspects of land use change, which apparently tightened control over urban development ('traditional localities'). At the same time, comprehensive town planning received widespread support. One result of this continuing growth was urban crowd and sprawl, which led to the broadening of planning concerns and adoption of rational planning techniques in urban policy ('suburban localities'). The weak integration of urban planning with economic strategies and other urban policy meant that urban planning's most important role was in reshaping attitudes towards growth. In the 1970s and 1980s the continuous growth of urban

fringe indicated failures to achieve a coordinated approach towards land use issues ('dispersed localities'). At the end of 1980s
the 'flexible thinking' of urban restructuring meant that urban
planning lost its strategic approach, became increasingly localised
and operated with *ad hoc* planning instruments and techniques
('renovated localities').

With the crisis of modernism, the erosion of nation-state functions, and especially the *re-evaluation of the welfare state* there has
arisen an important (mega)trend in contemporary (and future) urban development. In a more generalized sense, the socio-spatial
paradigm of modernization and welfare state formation included
(cf. Cooke, 1988, p. 483): (1) the relatively even geographical spread
to the periphery, semi-periphery and suburbs of modernization
processes, (2) the partial convergence in income and unemployment indices between the classes and the regions, (3) a characteristic posture towards producing standardized products for volume
markets, aiming for economies of scale, (4) economic development
in close proximity to new 'collective consumption' environments
(cf. CBD areas and strong city centre development), and (5) a characteristic demand for skilled and semi-skilled labour to fill secure,
'lifetime' occupations.

Figure 4.4 Spacing - scales of urban restructuring

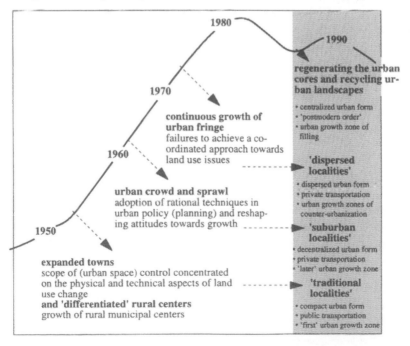

   Postmodernization has been constructed as both an ideology and a series of practices arising out of the failure of modernization to regenerate the economy of Western societies. The principles underlying postmodernization include: the appeal to unadorned market relations as a source of economic success, the rolling back of the welfare state, the weakening of established solidarities in the workplace, and the elevation of the private over the public sphere in matters of cultural and social provision. The socio-spatial form of postmodernization and dismantling welfare state consists of: (1) a markedly uneven spread of postmodernization characteristics (the emergence of regional differencies - North-South divisions, not only at national but also at international level), (2) the empirical divergence of income and unemployment indices between classes and regions ('A'-class people and 'B'-class people), (3) production disposed towards customized output, for niche markets aimed at economies of scale (increased product differentiation such as specialized malls, theme parks, power centers), (4) economic development occurring in areas of privatized consumption (shopping malls), and (5) limited and insecure labour market opportunities (part-time work and informal activities).

   The pressure to reorganize the interior space of the city has been considerable under conditions of postmodern principles. The vitality of the central city core has been reemphasised, themes such as the quality of urban living (gentrification, consumption palaces and sophisticated entertainment) and enhanced social control over both public and private spaces within the city have assumed widespread importance. During this postmodernization process, 'public-private partnership' of today amounts to a subsidy for affluent consumers, corporations and powerful command functions to remain in town at the expense of local community needs. Urban governments have been forced into innovation and investment to make their cities more attractive as consumer and cultural centers. Such innovations and investments have quickly been imitated elsewhere. Inter-urban competition has thus generated leapfrogging urban innovations in life-styles, cultural forms, products, and even political and consumer-based innovation. Herein lies part of the secret of the passage to postmodernity in urban culture (Harvey, 1987, p. 265).

Figure 4.5 Timing biases of urban policy

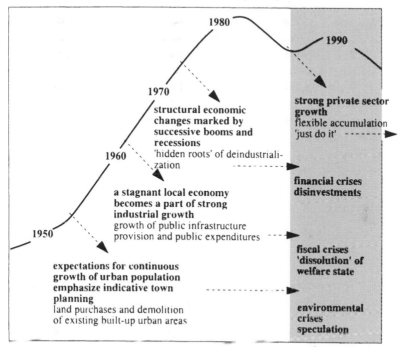

Markets, overconsumption and 'hyperspaces' are characteristic of postmodern urban form. According to Zukin (1988, p. 435) postmodernization refers to the structural polarity between markets and places, between the forces that detach people from or anchor them to specific spaces. Postmodernization also refers to the institutional polarity between the public and private use of urban space, and emphasizes markets over places. In another discussion Zukin uses the concept of *symbolic economy* to point out how entrepreneurial activity is becoming a powerful means to control cities (1995, pp. 1-11). In the 1970s and 1980s, the symbolic economy rose to prominence against the backdrop of industrial decline and financial speculation. The entrepreneurial edge of the economy shifted towards deal-making and selling investments. The growth of the symbolic economy in finance, media, and entertainment may not change the way entrepreneurs do business. However, it has already forced the growth of towns and cities, created a vast new work force, and changed the way consumers and employees think (here *place promotion policy*).

During the last twenty years, *cultural policy* has become an increasingly significant component of economic and physical regeneration strategies in many European cities. In terms of strategic objectives of cultural policy, the most important historical trend is the shift from the social and political concerns prevailing during the 1970s to the economic development and urban regeneration priorities of the 1980s. During the last decade, a shift to the right in the political climate in most West European countries and growing pressures on the financial resources of global government helped to downgrade the earlier emphasis on the importance of access to culture. It also undermined the view of culture as a contested political issue and of cultural policy as an alternative to the traditional strategies for political communication and mobilisation.

The strategies of the 1980s emphasised political consensus, the importance of partnership between business and public sector agencies, the value of 'flagship' cultural projects in promoting a city's image and the contribution of culture to economic development. In fact, many casestudies of European cities reveal that the direct impact of 1980s' cultural policies on the generation of employment and wealth was relatively modest when seen in comparison with the role of culture in construction positive urban images, developing the tourism industry, attracting investment, and strengthening the competitive position of cities (Bianchini, 1993, p. 2).

The cultural sector has an important economic role in future development of cities. *Cultural industries* feed both products and innovative ideas throughout an economy. In many urban renewal processes, cultural policy has also meant changes in power structures. As the initiators of structural changes, private developers and cultural experts (producers of urban land) have been the true architects of the urban spatial structure and urban landscape. Even though the developers operate in an unstable system consisting of the value judgements of private individuals, and under increasing public scrutinity, they are still able to assume the dominant role, since they are producers of the majority of the new buildings which households and businesses come to occupy (cf. Bourne, 1976, p. 539).

David Harvey (1985, pp. 3-7) characterizes the economic process whereby consumption has become dominant over production as a movement of capital away from the primary circuit (capital flows in the production process) to the secondary circuit (involving the capital which flows into fixed assets in the built environment and into the consumption fund). The gradual erosion of the 'industrial hegemony' is encouraged by a number of circumstances which arise periodically within the process of de-industrialization (overproduction, falling rates of industrial profit, lack of investment opportunities, etc.). Instead of traditional investments in industrial growth, these factors lead to 'flexible accumulation', when too much capital is produced relative to the available opportunities to employ it. Financial institutions and estate investors seeking to invest their considerable pools of capital in projects with maximal returns, invest more readily in the built environment under conditions of flexible accumulation (cf. Tweedale, 1988, p. 189). In periods of general prosperity, redevelopment of the built environment is a lucrative form of investment and property speculation.

There have been a number of renewal projects in Finnish cities since the mid-1980s, mainly concerning the intensification of land use in city centres and the reuse of old inner city industrial areas. A common feature of all these projects has been the unsystematic nature of the land use agreements reached between the property owners and the local authorities. A typical solution is that property owners submit to conditions which impose more financial burdens on them than under the regulations of the Building Act in exchange for greater permitted building volume, which will increase the value of the property. Land use agreements usually contain various stipulations which allow local authorities to influence the implementation of the land use plans. Nevertheless, it has been the practice that quite extensive responsibility and latitude in the renewal process has been delegated from the public to the private sector. This could lead to excessive sameness and expense in urban renewal, and more importantly, to a slackening of local government responsibility for the urban landscape. In most cases of redevelopment, local politicians have to learn a new way of managing land use policy. Traditional policy was connected with the supply of raw land, and usually took the form of contractual agreements between city authorities and building companies (developers). In the recycling of urban space, however, the private sector is more fragmented, and local government is actually uncertain as to what it is supplying and to whom.

Finnish cities (excluding perhaps Helsinki) are not large enough to create growth areas outside traditional urban realms. Instead of outer urban growth, the most 'promising' growth areas in Finnish cities are inner city edge areas - former industrial, warehouse and railroad zones. In many cases, these areas have been transformed into new kind of urban spaces - technological, ideological, consumption and financial - and they also have important role in inner growth of the city.

A major influence of political change on urban development are the broad ideological swings and shifts that occur from time to time in *urban planning and urban policy*. One question which emerges concerns the relationship between the 'private sector' city and the 'public sector' city. Results of recent urban policy show that 'public-private partnership' has emerged or is emerging as a leading instrument for urban development, especially in the reusing or recycling of urban areas. This has provoked extensive debate about new styles of urban planning. In particular, this is likely to stimulate a re-evaluation of urban planning concerning: (1) measurement factors of planning, such as changes in population development, changes in industrial structure and the dismantling of the public sector; (2) the dispersal of urban structure - including the new developments of urban fringe areas; (3) the integration and intensification of urban structure, including a reassessment of the significance of master plans in the reuse and 'filling' of existing built-up areas; (4) the public-private partnership policy; (5) the contradictory goals of urban development, such as simultaneously steering urban growth into inner cities and into fringe areas; (6) the control and channelling of competition - such as that between the inner city, competing suburban centres and areas of different land use; and (7) the urban landscape, including the structural polarity between the factors of market and place, and the institutional polarity between public and private use of urban space. The role of the market in creating and managing the infrastructure of the cities is being reconsidered in many sectors ('new privatism').

Dispersal versus inner growth

There are many aspects of technological and economic development in the near future which suggest that the dispersal of urban structures will continue. Both transportation and information technology will increase the mobility of people and of functions, i.e. we will see a reduction in the significance of physical location and distance. This dispersal is also favoured by the typical Finnish ideal of living in individual, dispersed houses. Nevertheless, this trend is likely to be restricted by the declining capacity of the public sector to construct and maintain an expensive technical infrastructure, mainly communications and electric power networks. Is urban growth in terms of population or economic activity a necessary prerequisite for dispersal, or can growth proceed without dispersal (cf. the idea of ecological city)? What part is likely to be played in this trend by the high-rise residential suburbs of the 1960s and 1970s and other urban fringe areas? To what extent will the developmental forces exert differentiated impacts upon cities in different parts of the country or, more importantly, in different parts of the cities, e.g. their centres and peripheries?

Projects investigating the future development of urban structure have revealed many conflicting trends inside the city region. These include the uneven developmental stages and different roles of suburban centres (e.g., the stages of dormitory communities, independent regional centres, catalytic growth and dispersal and high technology); the restrictions upon urban fringe development (the limits imposed by the physical and social structure of the 'city, by zoning costs, by mobility, by accessibility, and by urban landscape), the opportunities facing inner cities (the potential reuse of urban built-up areas, and new land use policy), and re-evaluation of the relationship between inner cities and urban fringe.

In Finland, as elsewhere, functional decentralization is leading to more complex and heterogeneous cities and urban areas consisting of many competing centres (see fig. 6). 'Postmodern urban realities' are finding their concrete counterparts in new suburban centres which are very attractive locations for networking firms, and where the huge growth in jobs and other truly urban functions is likely to be concentrated. The new urban realms are developing around various kinds of suburban-downtown areas, e.g. existing regional centres in the city region, regional shopping malls, and around multipurpose office and specialty concentrations (such as airports and their surroundings).

Figure 4.6   The coordinates of urban form and spatial reorganization (modified after Bourne, 1991)

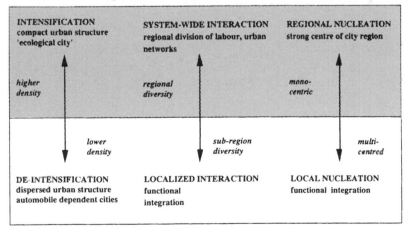

Discussions about alternative urban forms will raise questions which either promote centralization or dezentralization of urban space. Is it feasible to design alternative urban forms that offer a mixture of the positive features of the dispersed and compact urban forms? What conditions and factors encourage further reuse,

re-development, and intensification? At least four lines of argument are applicable here (cf. Bourne, 1991). One is based on conventional economic theory relating to the impact of depreciation
allowances, relative rates of return, and intraurban land rent differ-
entials on locational decisions. This theory suggests that capital
investment will switch between older developed and new
greenfield sites depending on the relative profitability of each and
the investors' evaluation of the risks involved. A second argument
is based on employment, lifestyle, and demographic considerations. It asserts that as the proportion of small, childless, two-
earner, professional households increases, the greater will be the
demand for alternative housing types and for the convenience and
cultural amenities of central area and/or higher density living. A
third argument relates specifically to the reorganization of urban
forms and to the effects of increasing congestion on the costs and
difficulties of commuting from distant suburbs to workplaces in
the central area and inner suburbs. A fourth argument is essentially
a political economy argument in which central area commercial
redevelopment and revitalization are encouraged by the local state
as a deliberate strategy to 'recapture' the decaying central core. The
purpose of the recapture may be to increase the municipal tax base,
to expel welfare dependent populations, or to facilitate local economic development and civic improvement.

Airport edge cities

The term "edge city" was popularized by Washington Post journalist Joel Garreau. In his book Edge City - Life on the Frontier (1991) Garreau provides the following five-part definition of Edge City. Edge City is any place that: (1) has five million square feet or more of leasable office space (the workplace of the Information Age), (2) has 600 000 square feet or more leasable retail space, (3) has more jobs than bedrooms, (4) is perceived by the population as one place, and (5) was nothing like a "city" as recently as thirty years ago. The growth of edge cities on land near airports and freeway intersections has been very fast since 1980s. Many of them overshadow their downtown CBD in terms of office space, and they have also begun to compete directly with the CBD for highly specialized functions such as banking, accounting services, luxury hotels, new telecommunication services, and even retail sales. More importantly, they are closely integrated with their own urban realm, their success having been closely intertwined with the development of "new towns", and of office parks and airport environments. Airports are important gateways everywhere. Many of the activities around airports are international by nature. Developments of airports and the surroundings are closely related and are significantly influenced by international developments. Airport edge cities are also specialized land-use nodes (commercial centers) surrounded by office buildings, hotels, hi tech enterprises, and logistic companies. "The combination of the present is the automobile, the jet plane, and the computer - the result is Edge City." (Garreau 1991).

Airports have a strategic impact on regional development, which is why "the airport edge cities" have to be integrated as far as possible into the urban planning policy of the city. In many cities near large international airports, their competitive position is closely related to the position of the airport itself. Amsterdam Airport Schiphol is the main motor for economic development in the region. One of the key areas in the Schiphol plan is to develop the airport and the region into a major international business center. The business city plan enables international companies to set-up an office at the airport itself. These prime locations give companies the possibility to link the airport terminal, the various office complexes, hotels and shopping facilities as well as with the extensive rail and road infrastructure. In the same way, the airport city has to be integrated into the commercial strategies of the city itself and into those strategies that guide the spaces devoted to leisure. Airports are increasingly becoming shopping centres of considerable size and have various elements that make them comparable to parks. In a certain sense, the airport shops in Barcelona are the continuation of the quality shopping streets of the city center and have the same kind of market image.

Harri Andersson

References

ANDERSSON, Harri (1995) Recycling urban landscapes - beyond the power, in BRAUN O. GERHARD (Ed.) *Managing and Marketing of Urban Development and Urban Life* (Abhandlungen - Anthropogeographie Institut für Geographische Wissenschaften) Berlin: Freie Universität Berlin, Band 52, pp. 585–593.

BEAUREGARD, Robert A. (Ed.) (1989) "Economic Restructuring and Political Response". *Urban Affairs Annual Reviews* (vol 34) London: Sage Publications.

BIANCHINI, Franco (1993) Remaking European Cities: the role of cultural policies, in Franco BIANCHINI and Michael PARKINSON (Eds.) *Cultural Policy and Urban Regeneration*, pp. 1-20. Manchester: Manchester University Press.

BLUESTONE Barry and Bennett HARRISON (1982) *The Deindustrialization of America*. New York: Basic Books, Inc.

BOURNE, Larry S. (1976) Urban structure and land use decisions. *Annals of the Association of American Geographers*, vol. 66, pp. 531-547.

BOURNE, Larry S. (1991) Recycling urban systems and metropolitan areas: A geographical agenda for the 1990s and beyond, *Economic Geography*.

CASTELLS, Manuel (1992) European Cities, the Informational Society, and the Global Economy. Amsterdam: Centrum voor Grootstedelijk Onderzoek.

COOKE, Philip (1988) Modernity, postmodernity and the city. *Theory, Culture & Society* 5 (2-3), pp. 475-492.

GIDDENS, Anthony (1979) *Central Problems in Social Theory. Action, Structure and Contradiction in Social Analysis*. London: The Macmillan Press, Ltd.

GLICKMAN, Norman J. (1987) Cities and the international division of labor, in Michael Peter SMITH & Joe R. FEAGIN (Eds.) *The Capitalist City - Global Restructuring and Community Politics*, pp. 66-86. New York: Basil Blackwell.

HARVEY, David (1985) *The Urbanization of Capital*. Oxford: Basil Blackwell.

HARVEY, David (1987) Flexible accumulation through urbanization: Reflections on 'post- modernism' in the American city, *Antipode* 19, pp. 260-286.

HOLCOMB, Briavel H. and Robert A. BEAUREGARD (1981) *Revitalizing Cities*. State College, Penna: Association of American Geographers, Resource Publications in Geography.

KEITH, Michael and Steve PILE (Eds.) (1993) *Place and the Politics of Identity*. London: Routledge.

KNOX, Paul (1994) *Urbanization. An Introduction to Urban Geography*. Englewood Cliffs, N.J.: Prentice Hall.

LASH, Scott and John URRY (1987) *The End of Organized Capitalism*. Cambridge: Polity.

LOFTMAN, Patrick (1994) Mega-projects and economic development. Paper presented at the conference on Shaping Urban Future, Bristol 10-13 July 1994.

LOGAN, John R. and Harvey L. MOLOTCH (1987) *Urban Fortunes. The Political Economy of Place*. Berkeley: University of California Press.

NIEMINEN, Jarmo (1994) Deindustrialization, the welfare state, and the laid-off worker's manner of coping. Tampere case study, in John DOLING, Briitta KOSKIAHO and Seija VIRKKALA (Eds.) *Restructuring in Old Industrial Towns in Finland*: pp. 215-238.Tampere: University of Tampere. Department of Social Policy and Social Work (Research Reports Serie A, 6).

PAHL, R.E. (1975) *Whose City?* Harmondsworth: Penguin Books.

PRED, Allan (1991) Spectacular articulations of modernity: the Stockholm Exhibition of 1897, *Geografiska Annaler* 73B:1, pp. 45-84.

TWEEDALE, Ian (1988) Waterfront redevelopment, economic restructuring and social impact, in B. S. HOYLE, D. A. PINDER and M. S. HUSAIN (Eds.): *Revitalising the Waterfront. International Dimensions of Dockland Redevelopment,* pp. 185-198. London: Belhaven Press.

ZUKIN, Sharon (1988) "The postmodern debate over urban form". *Theory, Culture & Society,* 5, pp. 431-446.

ZUKIN, Sharon (1995) *The Cultures of Cities.* New York: Blackwell Publishers.

5 Alternative Models for Spatial Urban Development: Policies for the City of Oslo and the Region of Viken

KARL OTTO ELLEFSEN

Introduction

The case study of the city of Oslo and the region of Viken discusses different alternative policies for large-scale spatial urban development. The study can be considered in terms of the different "worlds" of problems in urban development: (a) the empirically based understanding and conceptual definition of development trends in regional urban systems, (b) the discussion of ideals and models for urban development in relation to spatial, functional, cultural and political implications, and (c) alternative ambitions for governmental influence and alternative policies.

The alternative policies are discussed in relation to three general models for urban spatial development. These models are not only theoretical. They also represent, at least in terms of the actual case study, fundamentally different possibilities for action. The three models are:

1. the *compact city* model of urban concentration and regeneration;
2. the *città diffusa* model implying furthers deurbanisation, decentralisation and a polycentric urban pattern with a strong relationship to prevailing landscape qualities;

3. a integrated *urban expansion* model implying a clear geographi-
    cal definition of urban growth in areas strongly linked together
    by excessive mobility and advanced technical infrastructure.

These three models are evaluated as policies for future urban
development in relation to the city of Oslo and the surrounding
Viken region.

The main goal of this study is to relate spatial alternatives to
differences in ideologies and political priorities, and to point out
strong and weak aspects of the different strategies.

Figure 5.1 Oslo, the central agglomeration

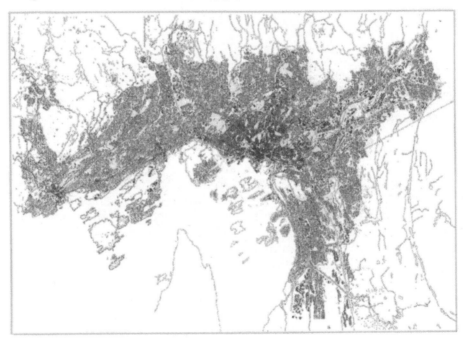

What is new, and what is not new in urban structures

In the urban transformation process now taking place in parts of
the European urban complex, it is possible to define various, and
often conflicting *development trends*. On the one hand there is a
strong tendency towards reurbanisation and the reestablishment of
traditional urban ways of living. This is most clearly indicated in
the gentrification processes that occur in most European urban
centres and is also reflected in dominating policies and cultural
tendencies. On the other hand, this tendency towards concentration is counteracted by a continuing process of regionalisation, ex-

tending the functional urban area, integrating different cities and areas into urban regional systems, and increasing the functional need for mobility. These counteracting tendencies are seen both in different coexistent preferences of ways of living and in differences in preferred spatial localisation for industrial development.

A few simple facts that characterise urban spatial development are first, that the city, functionally speaking, is expanding. Second, morphologically speaking, the city is no longer a continuous spatial system, but is constituted by a system of nodes, fields and areas with different functional and architectural characteristics. And third, the main economic, functional, social and cultural differences are not defined by the borders between the urban and the rural, but are to be found between economic, social, cultural and ethnic segments within the urban complex (Østerberg, 1998; Ellefsen, 1996; Secchi, 1993).

Both politically and professionally there are exist different approaches to *the concept of the city.* The family of concepts related to the "urban", depending of the scientific discipline they come out of, are based on morphological and architectural qualities, on cultural, social and socio-economic characteristics, or derive from institutional and administrative categories. The definitions trying to grasp the actual situation relate this family of concepts to diversity, possibilities for cultural, economical and social exchange, functional concentration and intensity and morphological density. The general trend is that all organisation of human settlement is influenced by, transformed into and characterised by an urbanised production and way of life. In this process, the traditional differences between the urban and the rural break down, leading to a situation where traditional models for understanding urban problems, and traditional ideals for urban architectural development are no longer valid. This situation is met by two different approaches. The first opinion states (Østerberg, 1998) that the family of concepts related to the "urban" no longer describes reality. The concepts refer to traditional ideals and are to be looked upon as relics from the past. Trying to liberate our minds from historical views that are passed on to us, we should therefore avoid "the city" as a concept and describe the characteristics of different concentrations. If the trend is that everything, or nothing, depending on what significance we give to the concept, is urban, we should change our categories to include an understanding that generously differentiates between many types of geographical and architectural answers to the new relationship between society and spatial organisation. Another opinion differentiates (Borja and Castells, 1997) between urbanisation and the city. This distinction offers an ideological meaning. It is fully possible to describe a future Europe composed of large, diffuse agglomerations. Industrial functions and areas for housing would be organised mostly along lines for transportation, surrounded by semi-rural uncontrolled areas for the poor and supplied by a set of service nodes according to market potentials. In this system, the elites will establish their own centres integrated into a global network for exchange of technology, production and finance. Their clearly bounded and controlled habitats will be adapted to different market segments according to income, social rank, interests and ethnicity. These kinds of dystopias are no longer only to be considered as science fiction.

The city defined mainly as a political and administrative system, represents an alternative to this dystopia. The characteristics of the city will still imply functional concentration and diversity and morphological density. And the city will display a relative urbanness[1] that can be measured with indicators that go beyond urban phenomena related to functional intensity and spatial density, and simultaneously encompass both territory and networks. However, a main characteristic of "the city" will be institutional: "by urbanisation is meant the spatial articulation, whether continuous or discontinuous, of inhabitants and activities. The city, on the other hand, both in the tradition of urban sociology and in the consciousness of citizens the world over, implies a specific system of social relation, of culture, and in particular of political institutions and self- government" (Borja and Castells, 1997, p. 2).[2]

This argumentation points to *the role of the city* in the future European society. The essential argument for sticking to the concept of the city, apart from academic requirements for precise categorisation, is that the city is a much needed phenomenon in future social organisation. This is because as a political and administrative institution, the city may play several roles in the exchange between local and global processes (Bookchin, 1995; Borja and Castells, 1997; Ellefsen, 1997). First, the city is a moderator and eventually a force for resisting global processes. Second, the city as an institution is a competitive actor improving, representing and selling local qualities. Third, the city represents a social section through society. Different social segments co-exist in the same geographical setting, in

---

[1]   The concept "urbanness" refers to Jacques Levy "Measuring Urbanness" in this book. Levy's intention is to update the concept of "the city" to be a descriptive concept for the existing situation by including new indicators.

[2]   This point, the meaning of urbanisation, is discussed in many of the articles, and from different points of view, in the books appearing in connection with the European Cost-Civitas program. "The specific system of social relation" referred to by Borja and Castells is illustrated by Dominique Joye and Anne Compagnon in their chapter on "Public Places and Urbanness" in this book.

co-operation and visible to everybody, and this setting is a means of avoiding socio-cultural disintegration. The city is still open to *the stranger*. And finally, the city is the institution where democracy can be further developed and policies discussed not only as ideology, principals and symbols, but also in terms of actual and very specific consequences.

The belief in the city as an institution represents a new focus in the professional discussion over *ideals for urban development*. Recent decades have seen both changing architectural or spatial ideals for urban structure and the tendency to focus on urban problems as the essentials. Following the intensive professional critique of modernistic urban ideals and synoptic planning practice, the focus was placed on the revival of traditional urban form. The ideals of the post-modern movement in architecture and urban design, dominant in the late 1970s and 1980s, propelled these ideas, which in some countries were even labelled as a "back to the city movement". In a European context, the focus, during the 1980s, shifted to the discussion of regional integration and urban regions. This overall spatial discussion has a counterpart in the interest for the potentials of the disintegrated, fragmented and nodal cityscape. In some respects, the term "urban region" could replace the term "city". The main interests of development policy was to exploit the potentials for integration in the urban regions by developing technical infrastructure, development corridors and new expansion in strategic regional nodes. The European urban policy of the 1990s, formulated in documents such as *Europe 2000+* (CE, DGXVI, 1994) and *European Spatial Development Perspective* (ESDP), (CE, DGXVI, 1997), may be looked upon as a recognition of the existing European urban pattern. Interest is directed more to discussing existing cities in new networks, making it possible for cities to develop different roles and cultivate their own potentials.

In countries where the Government has historically exhibited a strong political role in regulating and influencing urban development, a major discussion is also related to the *limitations and potentials of political influence on urban development*. Generally speaking, all major development strategies on an overall urban scale have presupposed extended governmental participation, public policies, investments and regulations (Castells and Hall, 1994, p. 240). This process is traditionally covered by the term "public planning".[3] However the term may also represent a relatively un-

---

3   Used here the term "planning" describes the general understanding of the concept up to the 1980s, involving different development strategies focused on public control and public action. In the current discussion concerning the role of governments and municipalities in urban development, it is useful to make a distinction between urban governance, planning and management.

productive practice if it leads to a command-economy and minute regulation of urban spatial development on a large scale, aimed at agglomerating the city into a pre-designed utopia. The type of action, however, that "planning" represents; definition of collective intentions, comprehensive analysis of knowledge gathered from a wide set of disciplines, co-ordination of different interests and the decision of a framework for future action is a necessity in order to develop integrated urban policies. In observing urban growth and transformation during the last two decades, the role of the government in executing urban policies in most European countries has changed. In a situation of strong urban growth combined with a high rate of transformation in built up urban areas, the question of ambitions in urban governance is once again very crucial.

Two typologies of Urban ideals: *compact city* and *città diffusa*

In the professional community, there are very different conceptions of a future "ideal" city. On the one hand there are notions that the traditional city, spatially and architecturally speaking, can be reproduced. In its most nostalgic manifestations this means that not only urban form, but also the citie's role in the production process, the integration between work and housing, and the social life of the city can all be reproduced. Within the "New Urbanism" movement (Katz 1994, Ellefsen 1997), this ideology is illustrated both in the development of seemingly traditional villages, in principals for urban transformations and in plans for the development of new regional systems consisting of traditionally designed urban nodes. The reconstruction of the city according to historically developed principals is intended to be an instrument in the reconstruction of social content, the establishment of "community", and the strengthening of "public man" in "civic society".

The belief in the role of the compact and traditional urban environment may also assume other forms. The essentially urban phenomenon is in this context not linked to given architectural semantics. The "urban" represents a system of cultural and social relations that manifest urban life. Morphological density is needed as well as functional diversity and a framework of legislative and political systems. In this perspective, a city such as Barcelona is an illustrative model containing a strong urban culture, metropolitan

---

Urban Governance refers to policies for urban development. Urban Planning refers to tasks and instruments of comprehensive and thematic physical planning. Urban Management denotes the methodology and instruments of putting policies into action.

density, and functional diversity. At the same time, the city manifests formal complexity within a generous set of structural and architectural principals. Urban spatial and architectural transformation processes may be viewed as tools for handling new social needs and adapting to a changing technological and economic framework. The city, looked upon as an institution, may be understood within a selection of morphological and architectural systems that differs substantially from a traditional European model. Illustrating this are the cities of Hong Kong and Singapore which have been fully transformed during the last two decades.

In relation to the *compact city* model the fact that an ideal model can be transformed into different spatial patterns, is a very important observation. The *compactness* can be manifested in different ways and it is not only possible to translate the model into reconstruction of the traditional city. The *compact city* can also be understood as a set of different responses to a series of intentions. First the most important aim is to reduce energy consumption and pollution by reducing transportation needs. Second, the intention is to create a city with functional complexity and integration that also reduces the need for commuting. And a third intention is to develop the city as distinct social and political unit.[4]

The *città diffusa* model is based on a different understanding of urban development. It assumes that the urban environment throughout the world has been transformed into a situation fundamentally different from traditional urban ideals. The "concentrated dispersion" of settlements that has occurred in most West European urban systems is not only a further development of traditional urban sprawl. It has characteristics related to spatial fragmentation, a complex system of functional roles related to the different nodes in the system, and a very relative centrality pattern.[5] This qualitative transformation is linked to fundamental changes in technology, production, culture and social systems and is therefore more or less irrevocable. The new regional urban system is characterised by its functional integration, and it stretches far beyond urban established agglomerations. The urban system is linked together by production and transport systems. Forces of change operate at regional levels and architectural transformations are therefore mainly tied to regional, not local or sitespecific, forces. From this point of view, the city may be understood as a network consisting of different nodes, corridors and areas having specific func-

---

4   For references to the *compact city* model see Anne Skovbro, "The Compact City and Urban Quality" in this volume.

5   The *città diffusa* is discussed on a more general basis by Giuseppe Dematteis and Francesca Governa in their "Urban Form and Governance, the new Multi-Centered Urban Patterns" in this volume.

tional roles and architectural characteristics. Urban systems produced by regionalisation do not have very much in common with
the classical European City anymore.

Instead we have contemporary "classic" cities: fragmented,
complex and reflecting different historical conditions. In terms of
architecture, this "new" state has long since become the everyday
environment of many people zapping between different spatial
and temporal constellations without effort and without losing their
perceptual grounding or ability to orient themselves in the chaotic
metropolis. Different terms are used for describing the new urban
condition. In the Danish discussion, the suburban carpet around
Copenhagen has been described as *mellemlandet* – the land *in between* the city and the countryside consisting of an archipelago of
areas for urban expansion (Juel-Christiansen, 1985). In French, the
term *la ville sanse aggolmeration* has been used (Fortier, 1993) to describe *"a form of metropolitanism that is not accompanied by density and
mass and which is thereby pre-eminently modern"* (Taverne, 1994). *Ciudad dispersa* describes the same phenomenon in larger Spanish cities
(Monclus, 1998). The American *edge city* concept also describes urban development that has some of the same characteristics (Garreau, 1991).

Maps show the urban region of Milan distributed like a patchwork metropolis covering a vast geographical area. Along the
middle Adriatic coast, the cities have agglomerated into what Dematteis and Governa describe as an *"urban continuum"*. The Randstad was seen as a ring of independent, separate cities bordering
"Het Groene Hart" of the Netherlands (Hall, 1984). Now a new
layer overlaps Randstad. It consists of housing areas outside the
established cities, supermarkets, theme-parks, areas with summer
cottages, technical infrastructure and freight terminals making the
cities a continuous agglomeration (Ellefsen,1998).[6] The Italian term
città diffusa is very descriptive for visualising this urban condition.[7]

The urban pattern of *città diffusa* is historically a result of possibilities created by private car mobility. It can be understood as a
combination of individual rationality (Secchi, 1996) and the rationality created by decentralised governance (Ellefsen, 1996). The individual, the family and the small firm are exploiting the new pos-

---

6   Wim Ostendorf, "New Towns and Compact Cities: Urban Planning in the
       Netherlands between State and Market", in this volume.
7   The term *città diffusa* emerged in an Italian context with a historical morpho-
       logical condition very different from the Norwegian case discussed in this
       paper. Norway does not display the network of small cities and villages that
       form the embryos of the diffuse urban pattern around a city like Milan. The
       Nordic "citta diffusa" is maybe a more genuine product of recent transfor-
       mational forces.

sibilities for alternative localisation. The municipalities now compete to attract capital, jobs and urban development, much like private enterprises on a free market.

As an archteype, the *compact city* is an open answer to a series of intentions. The *città diffusa*, on the other hand, is an archetype in a different setting. First it can be considered as a description of an existing urban condition. Second, this urban condition is understood as an irrevocable reflection of new technological, economical and social conditions. Third, the *città diffusa* (as the *compact city* model) can also be viewed as an ideal and desirable model for urban development. Basically, this view is based on the following premises: (a) The traditional city is no longer a necessary condition for production and modern consumption. In a society that places priority on innovation and high turnover, the large numbers and capacity that the traditional city represents is not an indisputable resource. Large cities may be considered as costly, nonfunctional and ineffective mechanisms containing various indigestion problems. Generally, the needs that create urban morphological density and functional intensity, in the future to a larger degree can be handled by IT and mobility. Although the growth of the new technology has so far not limited the growth of physical mobility (Veltz, Pierre, 1996), dual interaction is often needed,[8] and this contributes to the dispersion of the urban pattern. (b) Principels for housing and production localities in the future may follow different territorial patterns. The home may come to be more important as a workplace, and time spent in the home will increase. Generally, this will lead to deurbanisation, decentralisation and an urban pattern with strong relationships to prevailing landscape qualities. (c) A modern ecological perspective will not maintain the antagonistic distinction between nature and culture defining clear borders between the city and the natural landscape. The city may be seen a series of systems overlapping and coexisting with the natural ecological systems. (d) The traditional hierarchical urban system based on territory may in the future be substituted both locally, in the different urban systems, and in a global context, via the organisation of areas and nodes as a changing non-hierarchic rhizomatic system (Deleuze and Guattari, 1988).

---

[8]   The international airport is both a transportation hub and a place for meeting.

A geographical definition of Oslo and Viken

The meaning of the concept "region" has been rendered unclear because of its use in various contexts. The original Latin meaning of the word is a naturally and culturally homogenous geographical area, district or landscape. It was used this way by French regional geographers and also by Geddes and Mumford. The basic principle of classical regional planning was co-ordination of resources and interests within the development process of a region. Today the concept is tied to large-scale political intentions often reinforced by striking geometrical analogies.

Cities everywhere are susceptible to roughly the same forces. Wide urban areas are become knit together by the development of infrastructure and industrial restructuring affecting local systems. Norwegian urban regions, seen in a European context, also have certain distinguishing marks: (a) The number of inhabitants is low in relation to the large area covered by the regions. (b) Regions are also comprised of sites having very different historical back-grounds and landscape situations. (c) There is ample space avail-able for development compared to any Central European location. (d) The majority of citizens have preferred to live close to nature, in one family houses with large gardens close to the sea or the moun-tains, preferably both. This has influenced spatial organisation. (e) Until the 1980s, the Norwegian social democratic system was seek-ing to control environmental transformation to a very high degree through a system of planning procedures. However, it was not efficient in dealing with planning on the regional level. During the 1990s regional change has been influenced to a high degree by gov-ernmental investment in infrastructure. The prosperous Norwegian oil/gas economy further strengthens the possibility for political influence on critical choices concerning urban development. (f) The Oslo Fjord region is the only urban region in Norway in a Euro-pean sense. It extends 200 km in north-south direction and has one-third of Norway's population. The region houses a substantial part of headquarters for larger firms operating in Norway and nearly all governmental institutions. Apart from Oslo the region contains 12 other towns and around 50 other communities with a thousand or more inhabitants. Towns in Norway are usually situated on the coast with smaller communities lying further up the fjords and valleys. The Oslo and Trondheim Fjord, constitute an exception: both cities are located further up the fjord – Oslo at the very end of it – and the city has a hinterland of a more complex economic basis and regional pattern. (g) Oslo and other Norwegian cities included, generally both in the city centre and in CBD areas, are characterised by low density and an integration between landscape and urban typology.

The City of Oslo

The City of Oslo and the Viken region may be defined differently in terms of geography. Within its municipality borders Oslo is inhabited by a population of half million people. The rise in population during the 1980s and 1990s came after a period where Oslo, like most European cities, had experienced suburbanisation, decay in inner city housing areas, and loss of population in its central city areas. The current growth in areas with a traditional urban morphology is an indication of a new wave of Norwegian migration to urban regions, and apparently also a change in preferred ways of life, and in housing preferences. The boundaries of the municipality of Oslo, where the city meets preserved and protected areas of nature, worshiped as elements in the traditions of the Norwegian way of life, can be observed only as a clearly visible borderline.[9]

The circumference of the medieval city of Oslo is buried under heavy technical infrastructure in the harbour areas of the city. The Baroque City of the seventeenth century is clearly marked by its octagonal street-pattern. The borders of the urban municipality of the 1930s are also readable in the broad avenues and urban patterns described in the master plan of 1929.[10]

---

[9]   Both the Norwegian national consciousness and the way of life are very much dependent on the relationship to nature and natural landscape. The tradition of "tur", that is hiking through the landscape on foot in summer and skiing in winter has been an indisputable part of Norwegian culture. The Markagrensen, the legally defined border between built-up areas and nature to the east and west of Oslo has therefore been taken more or less as a "fact".

[10]   In the preface to the master plan guiding city development, "From the City of Kristiania to Greater Oslo", the author and chief of town planning Harald Hals, somehow with a feeling of resignation points out that "the plan is limited to the area that is legally the city of Oslo" (Hals 1929). The analysis and description, however "at least in the theoretical parts, inevitably had to be done regarding the natural urban body, which partly stretches far beyond the borderlines of the city".

Figure 5.2 Oslo, central urban area (Photo: Fjellanger/Widerøe)

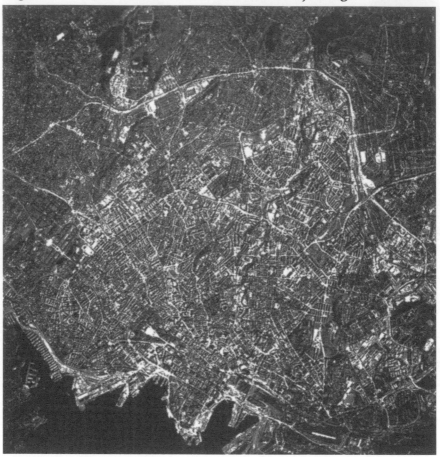

The city multiplied its territory when the two municipalities, Oslo and neighbouring Aker, joined forces in 1948. Urban development had already for a long time been taking place in areas outside the formal city of Oslo, often as a result of investment in private tram lines combined with property speculation, that made areas for the building of family villas readily accessible. This was especially the case in areas to the west and south-west of the city. The large urban expansion of the 1950s and 1960s therefore came to take place as an eastern extension of the urban area. In the history of Oslo the process is described as "a little town that exploded". The master plan for the growth of Oslo in the postwar years, like the plans for Stockholm, Gothenburg, in Copenhagen and in Helsinki, represent further developments of the planning principals established in the 1930s. Fritz Schumacher's plan for urban development along transport lines connecting to the central city of Hamburg, together with Abercrombie's plan for London from 1943, are the basic models for all these plans. The Oslo plan can definitely by viewed as what Peter Hall (1974, p. 58) called "a deliberate piece of social engineering". The satellite towns were divided into neighbourhood units organised around community centres. The idea was to connect territorial and spatial planning to the ideas of social control, and to help people to establish a sense of belonging to the place and to the local community. The social differences between the eastern and the western parts of the city had been established during industrialisation and were further strengthened by the planning of the 1950s and 1960s. While eastern areas were developed according to universal models for modern working class housing, the west was developed as private enterprise, in a patchwork of small housing projects.

The Oslo-agglomeration: "the natural urban body"

The agglomeration stretches far beyond the city limits. Topography, regulation, and transport lines constitute this "natural urban body". The topography set the path for the transport corridors as regulation obstructed the agglomeration from penetrating what was considered essential recreational areas. The steadily improving quality of the transport lines and the general tendency towards increased mobility, have led to prolongation of all three main corridors.

In terms of land-values and marked prices for building space, the western agglomeration is the most attractive both for housing and commercial use. The corridor has been termed "the consultant belt", and has indeed become the site of a large number of firms profiting from the Norwegian oil industry. Important factors for localisation were the vicinity to the former airport, and, in due time, the network established, but the determining factor was probably the proximity to housing areas preferred by the academic work force that the firms were competing for. Today the western corridor today extends about 30 km from the city centre.

The southeastern corridor has been the least attractive for both housing and industry, and the continuous agglomeration ends in more scattered development 15 – 20 km from the city centre. The building of a new highway and plans for a new railway-line through the area creates new possibilities for future land-use.

Figure 5.3 Oslo, the municipality (Sattelite-picture. Spot. Statens kartverk ®© CNES © Satellitbild). The area covered by Figure 2 is marked on the picture

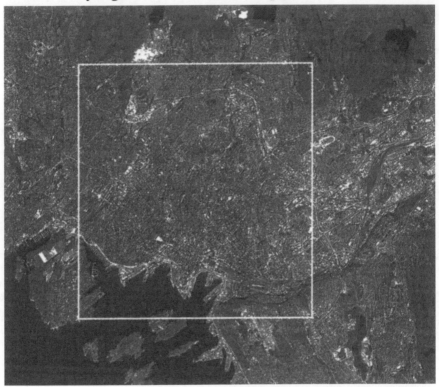

The largest and most important corridor related to industries is the northern. The industrialisation of Oslo in the post-war period was to a large degree concentrated in the Grorud valley to the north of the city. Today this area has the character of a "back yard" and partly of "waste-land" between industries, housing, areas for the handling of goods and other transport purposes. Architectur-ally speaking the area has the typical structure of the semi-periphery, each field having been developed following its own logical organising principles, and the system is knitted together by technical infrastructure and patches of left-over nature. Politically speaking, the northern direction for urban development is today, the most interesting. The new Norwegian international airport lies 60 km to the north of the city. The political intention behind this decision was to move the growth forces to the north and into the traditionally rural areas of the inland.[11] The agglomeration in the northern direction is wide and today extends about 30 km from the city centre.[12]

The population of the agglomeration is best described by the Norwegian statistical definition of the urban area (Myklebost 1960, p. 47–48).[13] A definition of Greater Oslo includes the central city and the 10 adjoining municipalities. The total population of this region is about 800 000 people. It has most of the characteristics of a functional urban region with a common labour market and housing market.

A definition that intends to grasp a wider functional region surrounding Oslo is more problematic. Here the discussion touches upon different degrees of functional integration. Oslo is the capital of Norway, and the city has historically developed important functional connections to the whole country. Traditionally, architectural or morphological criteria are of little use, the same may be said about sociological or cultural criteria for urbanity, or to political science criteria related to institutional, political-administrative boundaries. The matter is best handled in a context of economic or industrial geography, discussing the relativity of economic and functional integration, or even in a tradition of natural geography.

The first Norwegian understanding of the developing, functional urban region was published in the 1960s (Rasmussen, 1966, 1969) with studies of commuting and the drawing of maps of an area that could sustain commuters with a maximum travel distance to be covered within one hour. This understanding is further developed in later studies (Johnstad, 1996). The transport corridors, and the landscapes, towns and cities now in a process of functional integration, could be viewed as a wheel, with Oslo as the hub, and the transports corridors as spokes connecting outer areas to the hub. The population of this Oslo Fjord region is 1,25 millions inhabitants.

---

[11]   In the oil boom of Norway's economy, from 1970 onwards, the coastal areas have profitted while earlier industrial parts of the inland have had a relatively lower economic growth rate.

[12]   Existing growth-forces and policies of the municipalities to the north of Oslo indicate that the airport of Gardermoen, 60 km from the city-centre, will in a few years be integrated in the Oslo-agglomeration.

[13]   Norwegian statistics are very generous as to what is called "urban". The criteria for size is a minimum of 200 inhabitants. The morphological criteria is maximum of 50 meters between the houses. At least 75% of the working population should be occupied outside primary industries. A city is considered to belong to a larger urban agglomeration if at least one-third of the population is working in a central area of this agglomeration.

Figure 5.4   The Viken region. The borders of figure 2 are marked on the map (Landsat TM. ESRIN/Earthnet)

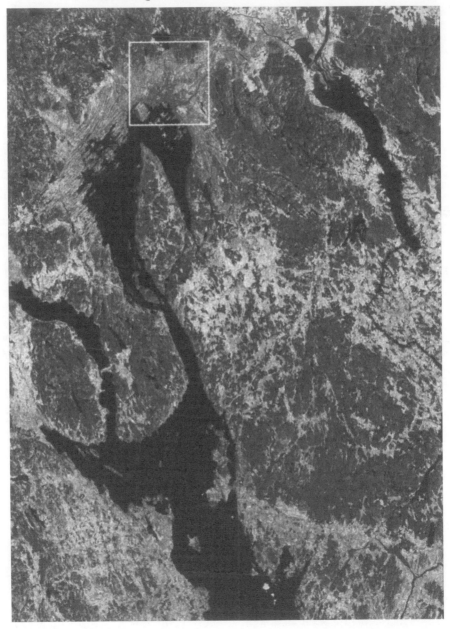

The Oslo Fjord region, Viken

Being the Norwegian part of the "Baltic Shield", Viken's land-scape differs in its geological characteristics from other parts of Norway, and this links "Viken" to the southern part of Scandina-via. The natural resources of the area, and geographical situation around the Oslo Fjord led to a historical development unique in a Nowegian context. Viken can be analysed as four types of ho-mogenous regions that were historically constituted by differences in geographical situation, in topography and in available resources (Tvilde, 1996). Each of these regions developed a specific economic and industrial structure, a role in a wider economic system, and a cultural specificity. The differences are still identifiable in the local dialects and in the historical architecture.

(a) The two coastal- regions[14] were bound to the fjord as a re-source and as a means of communication. The differences between the two regions are related to the proximity to Sweden in the east-ern coastal region, and to differences in potentials for industrial development. (b) The different inland regions are traditionally all rural areas and hinterland to the coastal cities. (c) The three river-regions have clear topographical borders, are centres of traditional communication structures, and the three urban agglomerations (Drammen, Fredrikstad/Sarpsborg and Skien/Porsgrund) are all major Norwegian industrial centres. (d) In this region Oslo has a distinguished role in its location at the end of the fjord, as capital, as the most important site for Norwegian industrialisation, and as the hub for all regional and national communication lines. Viken, seen as a landscape housing containing an urban functional re-gional system, includes 1.7 million people and is definitely the in-dustrial and economic centre of Norway.

Alternative models for spatial urban development applied in the discussion of the future development of Oslo/Viken

Transformation and urban growth

Both the city and the region are subject to strong transformational forces. (a) The regional integration process is transforming the Oslo Fjord region into a monocentric, functional unity. The process is facilitated by heavy government investment. The significance of Oslo as a communication hub has increased, and its dominance is

---

[14]   The coastal regions are Ytre Østfold and Ytre Vestfold. The inland regions are Indre Østfold, Indre Vestfold, Romerike and Ringerike/Hadeland. The towns are Drammen, Fredrikstad, Sarpsborg and Skien/Porsgrund (Grenland).

also strengthened by improvements in infrastructure. Weakening of the industrial base in other towns increases commuting to the centre. (b) The transformation of trade and industry has resulted in relocation and new functional roles, both in the regional system and for different urban areas within the city of Oslo. Traditional industries, the base of many towns, are now undergoing changes and are becoming impoverished. This new situation can alter the role of different cities within the regional pattern. The autonomy of towns has already dramatically decreased: they are losing their individual integration and identity, and their internal structures are being transformed from complex and comprehensive physical, economic and social networks to more specialised, monofunctional nodes. This means that local structural adjustments as well as measures to preserve local qualities are required. (c) Both the region as a whole and the city of Oslo are experiencing very strong growth in population. As in countries like England and the Netherlands, there is substantial housing demand, growing a need for 150 000 new units in Oslo and vicinity during the next 15 years.[15] (d) Compared to the other Scandinavian countries, Norway also has a large potential for urban growth.[16]

Along with general urban transformation processes and the need for political decisions related to major infrastructure developments, this provides a basis for a significant discussion of urban development models.

Città diffusa – applied

The model means that urban development may fully develop the positive and negative consequences of the decomposition of the city: if the trend is that everything, or nothing, depending on what significance we give to the concept, is urban. We should change our categories to include an understanding that generously differentiates between many types of geographical and architectural answers to the new relationship between society and spatial organisation.

Morphologically this model means that the city will disintegrate into a system of areas and nodes bound together by a strong system of communication. Viken should then be looked upon as a field of different nodes and condensations where various areas and nodes perform and specialise according to different roles, contain

---

[15]   The demand for housing is a result of (a) a rise in the total amount of households within the existing population, (b) increase in housing standards, (c) urban growth, and (d) immigration.

[16]   The industrial urbanisation process in Norway started very late and is out of phase with other West-European countries.

different functional profiles, exhibit different structural characteristics and reflect unique architectural characters.

Extreme *high mobility* is a necessary condition for this city to perform efficiently. Distance must be overcome with speedy communication systems in order to avoid indigestion. Daily meetings should be compensated by digital communication. *Urbanity*, understood as an open, dense and concentrated field of social contacts, must be cultivated in thematic zones for this purpose. Parts of the centre of Oslo will be further developed into "theme-parks", a performing "urbanity". Specific leisure areas might be developed in the same direction related to different seasons and lifestyles. So might contact-points to an international network such as the new airport city.

Within this system, *the economic structure* can develop. There are few needs for structural regulations. The most important criteria for localisation will still be accessibility within the regional system, agglomeration tendencies within different segments, possibilities for exposition and not least the selection of an environment that suits a selected business profile. The structure of business corridors and established nodes can live on, and market forces will handle the transformation. The capacity of the structure is related to the complexity, the many possibilities contained and the adaptability to dynamic change.

This model for urban development corresponds to a tendency where *the dwelling* gains more importance, both as a place for living and as a place for work. The localisation of the dwelling quarters can be even more separated from areas where modern production is concentrated. Preferences in ways of life, diverse ethnic groups and distinct social segments may "choose" the localisation of their dwelling within the wide field of different concentrations. And in this model it is an uncomplicated policy to locate new monofunctional dwelling areas in beautiful natural surroundings. As in the tradition of American *New Urbanism* and modern Dutch Housing, architecture can be given the task of developing the different areas as separate identities. Generally, the model considers the amount of space a sparsely populated land like Norway can offer in order to fulfil every dream of living with nature.

Ideologically, the model can be related to the notion of *Global Village*. Doxiadis, in the 1960s developed the idea of *Ecumenopolis* describing a worldwide, borderless urban system "which will permit the maximum variety of individual expression within a world-girdling system"(Doxiadis, 1960).[17] In the current incarnation of

---

17   Quoted from an interview with Doxiadis in the *Christian Science Monitor*, September 9, 1968.

this idea, the nomadic new man shall move in a non-hierarchic, rhizomatic and dynamic system like a fish in the worldwide continuous ocean. The model, however, also parallels traditional Norwegian regional policy with its intentions of regulating urban growth, maintaining the rural population and territorially dispersing regional growth.

Figure 5.5   The Oslo/Viken region

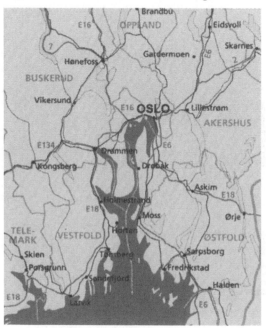

Examining *institutional borders* related to the political-administrative system, there will be no need for fundamental changes. The *role of urban policy and urban planning* will be related to the development of heavy infrastructure, co-ordination of the development of knowledge resources, conservation policies for natural and cultural resources, and activities to promote industrial development. The specific *architectural* challenge in this urban environment will be to give the nodes and the connecting structure a distinct architectural scenic character. The urban system as a whole is no carrier of a traditional urban language, and the esthetic experience is just as much based on the landscape and the relationship between the built and the natural landscape.

In the Oslo situation, this model is an expression of a trend scenario, of future reality if the prevailing urban development continues. It carries within few changes in existing politics, and a collection of planned development areas in the municipalities of the Viken region will be a tentative illustration of this future.

Compact city – applied

The model applied to Oslo implies that urban development is concentrated more or less within the existing agglomeration of Oslo. A key idea is that the city as an institution and the urban architectural principals carry a system of social and cultural relations that still have profound meaning, and that "urbanity" and urban life is based on morphological density, social and cultural intensity and functional manifold. In this context, the region of Viken is seen as a polycentric urban system of intensely developed urban units, with Oslo as the given centre of gravity.

The model has very clear *morphological* consequences. Political will and economic ability is very much needed because the model is not easily implemented and implies conflicts between different political intentions: (a) there is a contradiction between the need for transformation and established conservation policies; (b) the contradiction between established green structure and the need for expansion areas; (c) the conflict between high land-values in some of the open dwelling areas in the city and the eventual development of dense low-income "social housing".

Adjacency compensates needs for high regional mobility and the model is more than well suited for the further development of a collective transport system. The city in itself will be a concentrated urban and publically accessible field. This means that the model can stimulate the redevelopment of a *traditional urban commercial centre*.

The *structure of industries* represents two fundamentally different possibilities. One principle is that industries should be localised in the city centre. This might be a possibility, due to changes in the organisation of production. Expanding industries, also in Oslo, belong to the producer-services complex (Sassen, 1994). These mostly small firms, are easily integrated into the general urban structure and combined with housing. The second possibility is a reversal of the ideals for the modernist city and also very much in line with existing trends. The gentrification process leads to a central urban core dominated by housing, while industrial activities are located in the outer areas and along the corridors of transportation. Gentrification leads to conflicts with traditional industries in terms of traffic and noise. In Oslo at the moment in Oslo this leads to a self-generating process of de-industrialisation of central city areas.

The model may support the development of *modern urban ways of life*. Housing areas will have high densities and collective solutions, for different facilities must be chosen. A central discussion is what this modern way of life really consists of. A hypothesis could be that gentrification in Oslo, for people who can afford it, means a season-based nomadic condition moving between different dwellings: the summer house at the seaside, another in the mountains, a few weeks in the Caribbean with mobile office, and the high-priced modern flat in Oslo central city as a base-camp for further climbing. In the selection between urban and rural, Norway and the World, the choice is both.

Ideologically, the model expresses a strong belief in the future role of the city: the city as a framework around an urban way of life, the city as a political institution, a rich urban culture, functional complexity within a structural and architectural well-handled urban scene. However, it is important to emphasise that this does not necessarily mean the reconstruction of traditional urban form.

In terms of *urban governance, urban planning and urban management*, the model is very demanding. Strong regulations and firm and consistent urban policies are definitely needed. In terms of *architecture*, the model is equally challenging in relation to developing and selecting principals for densification as well as in terms of developing new housing types for urban living.

Urban expansion – applied

This model does not have the same archetypal character as the two other models, which in terms of morphological principals and policies can be seen as oppositions. The model can be viewed as a combination of two different development principals: (a) planned urban expansion and urbanisation of suburban areas connected to the central city, and (b) the development of a set of urban centres or nodes, each with a certain functional profile, specific architectural characteristics (scenographic zones), and strongly integrated into the urban system by an effective transport system.

The model, has historic *referents* back to prevailing models of the early post-war years, when Abercrombie's plan for London set the standard for urban planning. The most important difference is that while the model of the 1950s can be seen as one-dimensional, describing satellite towns, neighbourhood units, green belts and collective transport lines as the cure for all ills, a similar model of today must be inclusive and prescribe a set of different urban futures within the general model.

Also, this model will imply densification strategies within the city area. The difference is that the available, and awaiting, area for urban development is much larger. It this way, it is possible to implement substantial building activities, both economically speaking and in relation to housing qualities. In terms of prevailing trends in development, the model will be able to handle very different demands for localisation of industry and housing preferences. It very much adapts to a differentiated society, with different groups and sectors of industry launching and living their own preferences.

Within the new expanded city, travel may be very efficient. As was the case in the city expansion of the 1950s, land use in the outer urban areas can be co-ordinated, with the expansion of train- and subway systems.

As with the application of the *città diffusa* model, the expanded city will contain space for *a complex centre-structure*. The city centre can be further developed as a representative capital of Norway, with national institutions, culture, recreation and entertainment, also aimed at the further development of tourism. Urban development in outer areas can further exploit existing industrial structures in corridors and nodes.

In the 1960s, when Oslo underwent a major transformation with expansive programs for social housing, town planning followed a few dominating ideas. Both politically and professionally, there was agreement as to preferred ways of life, urban structure and architecture for housing. The satellite towns and urban development separated by green fingers were not only <u>an</u> answer, but also <u>the</u> answer. The new model for urban expansion must offer a *more pragmatic and market – based housing policy*. The model shall provide space for many different kinds of lifestyle, different housing patterns and typologies ranging from densification and transformation strategies in the traditional urban core of the city, to further development of nodes, restructuring of traditional villages within the new city borders, urbanisation of satellite towns and even also the development of large-scale development areas in landscape settings.

The model must be put into action by *manipulation of existing political and administrative borders* and the establishment of a new urban political administration. This unit can be established as a "county" within the Norwegian political system, and this urban county may contain a multitude of municipalities.

The most difficult implication is *the challenge to politics*. A very reflective, powerful and persistent policy is needed in order to consciously manipulate the urban system in the intended direction.

Otherwise, the output of trying to implement this model also will be the *città diffusa*.

A most decisive challenge to the model is to develop concepts for the transformation of a large integrated urban area. This kind of challenge has been met before: in the relocation of the medieval city into the pattern of a Baroque ideal city in the late 17th century, in the plans for the modernisation and expansion of the city from 1929, and through the plan for the city of the 1960s. A striking difference, however, is that urban development must today follow totally different concepts. The new city cannot be "gestalted" as a continuous set of urban spaces and conceived as a "Gesamtkunstwerk". It must be developed as a system of areas, nodes and structures following their own architectural principles and exhibiting different formal expressions.

Some notes on the evaluation of the models

Seen in relation to each other, the first model follows a path where the traditional city is transforming its political, administrative, functional, social and cultural role, and is replaced by a wide, functionally integrated field exhibiting different functional and architectural densities. The second model, on the contrary, is intended to concentrate growth in the traditional urban centre, and by this is trying to regenerate urban ways of life, urban culture and social, administrative and political institutions. The third model is a model for urban expansion, and it intends to create a strong international city in the Oslo/Viken area by expanding the city and by developing different "technopole" strategies (Castells and Hall, 1994).

(a) A first evaluative point is to stress that the universal discussion of models for large-scale spatial models for urban development should take into account regional differences. Each urban region is trying to maximise it own compatibility by developing characteristics related to territory and place. These characteristics differ from Central and Southern European urban settings, to technopoles selling Mediterranean climate (Halvorsen, 1996,1998) and to a Nordic and Norwegian setting. Norwegian compatibility in relation to knowledge resources is also based on the closeness, culturally and in terms of geography, between the modern city and nature. In this respect, compared to countries like The Netherlands and Denmark, a city like Oslo can offer optimal situations for dwelling even combined with proximity to an urban multicultural centre of European scale and character.

(b) The *città diffusa* model is normally regarded as an unwanted side-effect of modern urban development. The model is substantially criticised from a sustainability point of view, on the basis of effects on public economy, on the basis of social and cultural consequences and in terms of architectural quality. There are certain parameters in this discussion that are disputable. One is the future relationship between on the one hand living/housing and on the other hand the organisation of work and the amount of home-based working; living and working both as very integrated contexts and contexts with a very different geography. Politically speaking, the model, in the form we know of today, is very much a reflection of market forces, and as such, not a primary object for politics. However, the model may be cultivated in a transport/land-use context and in a morphological context relating built up areas to their natural surrounding. Specifically in the case of Oslo, it must be noted that the ideal is coherent with dominant political intentions of regional development and the principals of relative autonomy in urban development policies for the municipalities.

(c) The *compact city* model is the politically correct model that most cities seek to put into action.[18] There are different interpretations of the model, but as a general principle the model can be investigated in three ways: Does the model offer what is demanded of it? Is the model possible to achieve? In relation to democratic principles, to what extent can the model be politically implemented? (Breheny, 1992).[19] As to the question of veracity, most interests have been channelled into relating the model to the intentions of sustainability. Scientific investigations (Næss, 1994) with Oslo as a central empirical case, have shown that both energy consumption and pollution can be reduced by using "compact city" strategies.[20]

But this gain most probably depends on specific changes in ways of life and does not accord well with a multi-dwelling lifestyle or a substantial increase in long-distance travel. A central discussion of course is the relative importance of urban structure compared to general changes in consumption and lifestyles. As to the question of feasibility, it must be noted that in an Oslo context, the model represents a major shift in the developing trends and the existing housing market, not providing areas for low-density

---

[18]   The "Compact City" model applied to Oslo is a very centralised model which concentrates on transformation and densification in central city areas.

[19]   Professor Michael Breheny, University of Reading, presented these three questions in a lecture at Lund University, Sweden, March 1999.

[20]   The conclusions of the investigations by Peter Næss show that the ultimate city in terms of energy consumption is a pattern of medium-sized to small functionally integrated cities.

housing. This means that the model cannot be applied if governmental authorities are unwilling to put strong regulatory politics
with substantial economic effects into action. And this leads to an
emphasis on the third question. Assuming that concentration is
feasible, would it be acceptable to the people affected? In a sparsely
populated and landabundant country like Norway, to what degree
shall urban concentration strategies be used if they conflict with
individual and group specific preferences. The evaluation would
change if the advantages of density and a multitude of urban links
would permanently alter housing and living preferences.

(d) Going back to the role of the city and the differences between urbanisation and the city, the question is, "Which model
makes it possible for the city to perform its role: as a moderator of
global forces, as an actor in relation to development of industries
and knowledge, as a open socio-cultural setting and as a democratic institution?" First there are many examples of successful regions (Castells and Hall, 1994; Halvorsen 1996) implementing different kinds of technopole strategies. In a Norwegian context, the
political pursuit of economic activities is also organised as co-
operation between cities in an extended version of the Viken landscape.[21] This is also part of the European ESDP strategies and does
not really conflict with a *città diffusa* model. However, this model
conflicts with the needs to expand the city as an open socio-cultural
setting and probably also the needs to develop democracy.

(e) *Compact city* intentions can be achieved by different large
scale spatial strategies: strong concentration in the central city areas, development of a polycentric situation with many nodes knit
together by an efficient collective transport system, and by a strategy for urban expansion as described in "the third model". The
advantage of this model, both as a general *leitbild* for urban extension and as an input to the discussion of urban development, is its
inclusive character. The model can cope with the different needs
and preferences in a modern urban system, and with the very important intention to create a strong, unified city.

In the hierarchy of urban policies - Urban Governance, Urban
Planning and Urban Management - the model challenges Urban
Planning and the need to discuss Urban Policies for large-scale ur-
ban development as questions of ideology.

---

[21]   The co-operation is organised in "Østlandssamarbeidet" covering the eastern parts of southern Norway. The work is inspired by European ESDP
strategies.

Nomadic man

> *"Write to the nth power, N-1, write with slogans*
> *Form rhizomes and not roots, never plant!*
> *Don't sow, forage!*
> *Be neither a One nor a Many, but multiplicities!*
> *Form a line, never a point! Speed transforms the point into a line.*
> *Be fast, even while standing still.*
> *Line of chance, line of hips, line of flight.*
> *Don't arouse the General in yourself!*
> *Not an exact idea, but just as idea (Godard).*
> *Have short-term ideas.*
> *Make maps, not photographs, or drawings.*
> *Be the Pink Panther,*
> *and let your loves be like the wasp and the orchid,*
> *the cat and the baboon.*
> *As they sing of old man river:*
> *He don't plant` tatoes*
> *Don't plant cotton*
> *Them that plants them is soon forgotten*
> *But old man river he just keeps rollin` along."*
> *(Deleuze and Guattari, A Thousand Plateaus.)[22]*

In sociological critique of modern industrial society there was a tradition of making distinctions between "Gemeinschaft" versus "Gesellschaft" (Tönnies), "mechanical" versus "organic solidarity" (Durkheim) or "community" versus "alienation" (Marx/Nisbet). The basic argument was that technical and economic changes in society inevitably led to qualitative transformations in the conditions for urban life. Tönnies was the first theoretician to introduce the longing for "the lost Gemeinschaft", invoking the positive virtues of traditional rural society with solidarity, respect, concern and benevolence. "Gesellschaft" referred to the mechanical and relative solidarity between individuals in modern urban society. An intention of traditional urban planning throughout the twentieth century has been to re-establish community and thereby "the lost Gemeinschaft". Even the goals of Agenda 21 can be seen in this historical context.

The distinctions deal with the relationship between individual identity and local community or "place". The pursuit to re-establishment of communities within modern urban settings has been constantly confronted by arguments that modern metropolitan man is taking part in a complex set of social relations not

---

[22]   New York, Athlone Press, 1988.

limited by small-scale geographical borders. The pursuit of "community" is there-fore, so to say, of little value.

The concept of "nomadic man" can be judged in this historical context. The assertion is that information technology will fundamentally change society. Re-ferring to Deleuze and Guattari, our society is no longer relevantly described by traditional hierarchical models but should be comprehended as a flexible, chang-ing non-hierarchical rhizomatic system. One basic phenomenon in the culture of this society is that the relationship between individual and "place" is undergoing a fundamental change. "Nomadic" as an analogy refers to the positive possibili-ties that the new technology is creating. The longing for the lost Gemeinschaft is overcome by developments in information technology: the world of modern man in his high-tech tele-cottage is not limited anymore; all the information of the world is delivered to any location. Modern man is liberated from the limitations of place and distance. Globalisation of culture, language and production makes it possible to feel at home "wherever I lay my hat". The nomad is at home when he is travelling. Movement represents identity and safety and so does every point he visits during his travels. "Home" is a concept that not only deals with family roots but just as much with a set of rituals, personal rhythms and everyday rou-tines, and these needs can also be interpreted and fulfilled within a nomadic way of life. "Home" in the traditional meaning of the word may at most be reduced to a "base-camp" from where the global player can make his assault on the summits of the world.

The arguments against this understanding are, firstly, that the practices of the few, young, wealthy globetrotters are made into a norm, and secondly, that the meaning and relative stability of local culture, language and local identity is underestimated. Thirdly, processes of desurbanisation should be the result of this development. However, signs of real desurbanisation do not appear very convinc-ing. On the contrary, at the scale of urbanised regions, processes of concentration are still important. This can be understood by asking what modern man needs. Modern man is still a social human being, not only interested in technology, but also in social relations with other human beings. Yet man also needs protection and therefore neighbours in his vicinity; co-operation with these neighbours he endeavours to protect his residential environment. In other words, modern man still needs some sort of Gemeinschaft.

The importance of the "nomadic man" concept is that it questions the dominant political goal to recreate community, to "return to place" and under-line local culture. It imbues the historic antagonism between "community ideals" and "metropolitan ideals" with a contemporary interpretation.

Karl Otto Ellefsen and Wim Ostendorf

References

BOOKCHIN, Murray (1995) *From Urbanization to Cities – Towards a new Politics of Citizenship*. New York: Cassel.

BORJA, Jordi and CASTELLS, Manuel (1997) *Local and Global. Management of cities in the Information Age*. London: Earthscan Publications.

BREHENY, Michael (1992) "The contradictions of the compact City: a review", In BREHENY, M. *Sustainable Development and Urban Form*, 138 – 159. London: Pion Limited

BREHENY, Michael (1996) Centrists, decentrists and compromisers: "views on the future of urban form". In JENKS et al. Eds. *The Compact City – A Sustainable Urban Form*, 13 – 35. London: E and FN Spon.

CASTELLS, Manuel (1997) *The Power of Identity – The Information Age: Economy, Society and Culture* Vol. II, Malden, Mass: Blackwell.

CASTELLS, Manuel and HALL, Peter (1994, p.240) *Technopoles of the World – The Making of the 21ˢᵗ Century Industrial Complexes*, London: Routledge.

COMMISSION EUROPEENNE, DG XVI (1994) *Europe 2000+. Cooperation pour L'amenagement du territoire europeen*, Luxembourg: Office des Publications Officielles des Commnautes Europeennes.

COMMISSION EUROPEENNE, DG XVI (1997) *Schéma de développement de l'espace communautaire. Premier projet officiel présenté à la réunion informelle des Ministres responsables de l'aménagement du territoire des Etats membres de l'Union européenne*, Noordwijk, Luxembourg: Office des Publications Officielles des Communautes Europeennes.

DELEUZE, Gilles and GUATTARI, Felix (1988) *A Thousand Plateaus*. New York: The Athlone Press, 1988.

DOXIADIS, C.A (1960) Dynapolis, *Byen for fremtiden*, Oslo: Oslo Arkitektforening.

ELLEFSEN , Karl Otto (1996) "Vår tids omgivelser: Ikke by – ikke land", in Fløistad, G., Moe., K and Thiis-Evensen, T., *Christian Norberg-Schultz, et festskrift til 70-årsdagen*, Oslo: Norsk Arkitekturforlag.

ELLEFSEN, Karl Otto (1997) "Tilbake til framtida, The "New" Urbanism", in *Byggekunst* vol.3, Oslo.

ELLEFSEN, Karl Otto (1998) "What is new and what is not new in urban systems", in *Byggekunst* vol.7, Oslo.

ELLEFSEN, Karl Otto (1998) "Randstad",in *Byggekunst* vol.7, Oslo.

FORTIER, B (1993) "La citta senza agglomerazione", in *Casabella* vol.599, Milano.

GARREAU, J (1991) *Edge City, Life on the New Frontier*, New York: Doubleday.

HALL, Peter (1974) *Urban and regional planning*, London: Penguin, Harmondsworth.

HALL, Peter (1984) *World Cities*, London.

HALS, Harald (1929) *Fra Kristiania til Stor-Oslo. Et forslag til Generalplan for Oslo*, Oslo: Aschehoug & Co.

HALVORSEN, Knut (1996) *Hva kan vi lære av Motpellier?*, Oslo: NIBR-report No 2.

HALVORSEN and LACAVE (1998) *Innovative Systems in Urban Areas*, Oslo: NIBR-working paper.

JOHNSTAD, Tom (1996) *Næringslivets utviklingspotensialer i Oslo regionen*. Oslo: NIBR.

JUEL-CHRISTIANSEN, Carsten (1985) *Monument og Niche*. Copenhagen: Rhodos.

KATZ, Peter (1994)*The New Urbanism: Towards an Architecture of Community*, New York: McGraw-Hill, Inc.

MONCLUS F. J. (Ed) (1998) *La ciudad dispersa. Suburbanizacion y nuevas periferias*, Barcelona: Centre de Cultura Contemporània de Barcelona.

MYKLEBOST, Hallstein (1960) *Norges tettbygde steder*. Oslo: Universitetsforlaget.

NÆSS, Peter (1994) *Energibruk i 22 nordiske byer*. Oslo: NIBR.
RASMUSSEN, TOR Fredrik (1966) *Storbyutvikling og arbeidsreiser*. Oslo: NIBR.
RASMUSSEN, TOR Fredrik (1966) *Byregioner i Norge*. Oslo: NIBR.
SASSEN, Saskia (1994) *Cities in a World Economy*, California: Pine Forge Press.
SECCHI, B. (1993) "Le transformazioni dell'habitat urbano", in *Casabella*, vol.600. Milano.
SECCHI, B. (Ed) (1996) "Veneto" in Clementi A., Dematteis G. and Palermo P.C., (Eds) *Le forme del territorio italiano. Il Ambiente insedidativi e contesti locali*, Roma-Bari: Laterza.
TAVERNE, E. (1994) "The Randstad. Horizons of a diffuse city", *Archis*, vol.7, Rotterdam.
TVILDE, Dag (1994) "Regionale bysystemer", in *Byutvikling – tendenser, tolking, planlegging*. Oslo: AHO, Dep. of Urbanism.
ØSTERBERG, Dag (1998) *Arkitektur og sosiologi i Oslo – en sosiomateriell fortolkning*. Oslo: Pax Forlag.

Part II
Urban Qualities and Urban Life

6 The Compact City and Urban Quality

ANNE SKOVBRO

Introduction

One of the major challenges of urban planning is achieving a more sustainable urban development. This has been recognised not only by several governments, but also by the EC, the UN and by several other institutions such as Friends of the Earth. One of the strategies for a more sustainable urban development that has found support among many researchers, governments, the EC and NGOs, is the compact city strategy, a strategy implemented in national planning guidelines in several northern European countries such as Denmark, England, Norway and especially the Netherlands.

Although the actual content of a compact city strategy varies from country to country, the general idea of the compact city strategy is to locate future urban development within current urban boundaries. Thus, a large proportion of future urban development will take place through urban densification. The main arguments for such a policy have centred on resolving transport and transport related environmental problems. Research has documented that a more compact urban form uses less energy for transport, thus reducing the level of global and local emissions. Furthermore, it has been argued that more compact urban development reduces the loss of agricultural land for future urban development. However there is no widespread consensus about the benefits of a compact city strategy, as the following section will show.

The discussion has tended to focus on whether one is for or against the compact city concept. Yet there are several problems in the discussion. First, the compact city strategy is often discussed from different points of departure, typically with only one aspect of sustainability or one geographical scale in mind. Research has often focused on energy conservation, especially the relation between urban form and energy use for transport. However, there is more to sustainability than energy conservation. Furthermore, the debate over the compact city concept is rarely related to some of the current problems facing urban development and planning. Increasing globalization and the competition between urban regions have lead to new public-private partnerships in planning, and new dynamics in the urban area. How does a compact city strategy, with its goal of sustainability work together with these other trends? Finally, the debate implies that we must choose between a compact city development or a more widespread green urban development. However, the existing urban structure is very difficult to change, and realistic strategies for a more sustainable urban development must be based on the existing cities and towns and on an accurate picture of the environmental problems in these areas and regions. Danish towns and cities, have many former harbour areas, industrial sites and railroad areas that need to be redeveloped. While compact city development through densification may very well facilitate a more sustainable urban development in some cases, it will not necessarily be the universal solution. It would therefore be more valuable to discuss how to find a balance between the goal of limiting the use of resources on a larger scale, while maintaining a desirable urban environment on a smaller scale. In order to do so, we need to understand the relationship between the compact city and the quality of the urban environment.

This article will focus on the contradictions between the compact city and the quality of the urban environment. It comprises a general discussion of the relation between the compact city and the quality of the urban environment and a case study on the urban environmental consequences of the current densification of a central Copenhagen district. A discussion of the consequences of urban densification can serve to test the claims and counter claims on the compact city, and, more importantly, raise the issue of balance between urban densification and urban quality. Only by discussing this problem can urban densification be used as a tool for generating sustainable urban development.

Why a more compact city?

The compact city strategy is based on several arguments. The dominant argument has been related to energy conservation. Research has demonstrated a close correlation between high urban density and low energy use for transport (Newman and Kenworthy, 1989; Næss, 1994), and between high urban density or compact urban form and a low energy consumption for space heating. A more compact urban form support efficient district heating systems (Owens, 1992), while multi- storey housing has lower energy consumption pr. square meter for heating than detached single family housing (Næss et al, 1996). In addition to these energy saving arguments, the EC, in their Green paper on the urban environment, has argued that a more compact city will create a more liveable urban environment and that urban concentration will support local services (CEC, 1990). A final argument that has gained support in recent years is the "loss of land" argument. Continued urban sprawl will lead to a continuing loss of land, and often valuable agricultural land. This speaks in favour of a concentration of urban development within the current urban fabric. The Foundation for Protection of Rural England has been one of the major proponents of this position which is also supported by researchers involved with biodiversity issues. They have stressed that from a biodiversity point of view, large green areas outside the city are more valuable than many small green spots within the urban area (Nyhuus and Thorén, 1997).

The compact city strategy, as already mentioned, has also been criticised. The critique has focused partly on the relation between energy consumption and urban structure. Gordon and Richardson (1990) have pointed out that variation in petrol consumption the cities studied by Newman and Kenworthy, is due mainly to differences in lifestyle and travel behaviour (cited in Breheny, 1992). Others have mentioned that more compact city development causes increased traffic congestion, which leads to greater air pollution in urban areas. Nevertheless Breheny - a critic of the compact city strategy - has recognised that "even if the energy argument does not stand, the loss of land argument probably stands, and a continuing movement of people developing their own acre plots in the countryside is not an attractive proposition" (Breheny, 1996). Breheny and others have put forward a critique on the more basic concept itself; which picture of a sustainable city does the compact city concept produce? Is this the kind of sustainable city that people would choose to live in? When it is claimed that the compact city will foster social and cultural diversity and activity, and provide a livelier, safer, and socially equitable environment, the counter arguments are that higher densities lead to more crime, the disadvantaged will suffer more from the resulting high land prices, noise and pollution, and that the compact city is not socially acceptable due to perceptions of overcrowding and loss of privacy (Burton et al 1996). Thus, the critique has focused on the quality of the urban environment that a compact city strategy would produce and has questioned whether this urban environment will be seen from a local point of view as sustainable and desirable (Breheny, 1992; Thomas and Cousins, 1996). Breheny (1992) has stressed the fact that the dominating choice of life-style is suburban. Furthermore, Næss has conducted a survey that supports the resistance towards a densification of housing areas and a desire towards new residential developments in country-like surroundings (Næss and Engesæter, 1992). Hence, there is a conflict between densifying the urban fabric and the preferences of its inhabitants.

The contradiction between the compact city and a green city has also been a central subject in the discussion. The EC, in their document "Green Paper on the Urban Environment" (ECE, 1990), has advocated more compact urban development and at the same time for a greening the city. Environmental organisations such as Friends of the Earth have done the same thing, though the latter have recognised the conflict between more compact urban development and achieving a green city. The importance of green space in the city has been related not only to the recreational value (Breheny, 1992; Grahn, 1991; Thorén and Nyhuus, 1997) but also to the ecological needs (Thorén and Nyhuus, 1997; Orrskog and Snickars, 1992). Furthermore, Rådberg (1995) has stressed the need for green areas or space for recycling waste in urban areas. He claims that recycling of waste is difficult under conditions of high density urban development.

Beyond claims and counterclaims

How do we take the discussion and research about the compact city further than mere claims and counterclaims? In general, there has been a lack of empirical documentation, especially on the local implications of a compact city strategy. Most of the empirical research has had a regional focus, often with the transport/energy use discussion as the predominant topic. In recent years, empirical research has led to a discussion about the local consequences of more compact urban development (Guttu et al., 1997; Burton et al., 1996). The case study to be presented here is a contribution to this effort.

The critique of the compact city concept is certainly relevant. Viewed in terms of sustainability, we observe a part of the critique focuses mainly on local aspects of sustainability, while another aspect focuses on more global aspects. Following the Brundtland report, sustainability is defined as a development that meets present needs without compromising the ability of future generations to meet their needs and aspirations. Many critics have asserted that this is a very wide definition, but most agree that sustainability concerns more than decreasing our use of resources, it also involves wider environmental issues and social, economic and cultural issues.

We may conclude that a study of the sustainability of an urban area includes several levels of sustainability and several topic's that need to be elucidated. Owens has discussed this in terms of the internal and external sustainability of the city (Owens, 1995), but the issue could also be understood as several geographical levels of sustainability: regional, urban and local. Kevin Lynch in his normative theory of good city form, has stressed the importance of social and democratic issues (Lynch 1985). Many aspects of the critique of the compact city concept are thus relevant, especially the need to gain empirical knowledge as to the actual consequences of densification. In the subsequent case study of the urban environmental consequences of densification in a Copenhagen district, the following issues have been assessed in terms of a definition of the urban environment which includes physical and structural urban issues, the use of resources (especially for transport), the quality of the environment (air pollution, noise, traffic accidents, cultural heritage, access to green areas, urban space and services), and socioeconomic issues (jobs, services and social segregation). This understanding of the urban environment is developed partly through the EC Greenpaper on the Urban Environment, but also through the concept of sustainable development, which not only entails a reduction of the use of natural resources, but also socio-economic issues such as democracy and access to services. Through these issues, the critique, claims and counterclaims, in relation to the compact city debate, will be addressed. In relation to the transport debate however, the focus of this article will be on local issues. The regional aspects have been thoroughly discussed in other papers.

A case study of urban densification in Copenhagen

The Copenhagen district of Østerbro has in the past decade experienced the highest level of building activity of all Copenhagen districts. The district, located next to the old central district of Copenhagen, was built during the period of 1880 - 1920. It contains three main areas: a housing district of 5-6 storey closed blocks, interrupted by small spots of remaining detached houses or villas; a park area where public facilities such as hospital, university and

sports stadium have been located; and the port area where a substantial redevelopment is currently taking place.

**Figure 6.1   Map of Østerbro and case areas. Part of map with
reference to agreement G18/1997 with The National
Survey and Cadastre**

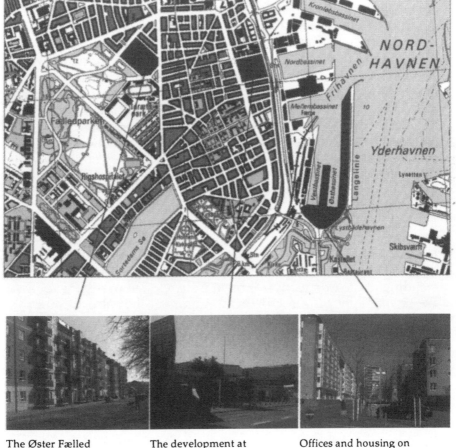

| The Øster Fælled | The development at | Offices and housing on |
| development | Kryolitgrunden | port areas |

Normally, the district is perceived as a "low density" area
compared to other central Copenhagen districts. This is confirmed
by the official statistics. However, the calculation of the density of
the district includes port areas, roads and green areas. If these are
excluded the density of the residential district reveals itself to be at
the same level as other central districts. Additionally, employment
density could be included to obtain a more accurate picture of the
human density of the area. Fouchier (1998) has argued that an ac-

curate picture of the human density is vital in order to discuss densification of urban areas.

In the district of Østerbro, 76.000 square meters of housing and 150.000 square meters of offices and retailing has been built during the 1990s. The main reason for this high level of urban restructuring is the release of large Copenhagen port sites, former ministry of the defence and industrial sites to private investment. Redevelopment of these areas is of major importance and interest in assessing the consequences of densification in this case study. In addition, the local authorities have areas which they would like to redevelop. Several of the sites are located within a 500 meters distance of the three metro stations of the district. Thus, localisation and high density development on these sites is in line with regional and national guidelines, stating that new commercial developments are to be localised no further than 500 m from a transport junction such as a metro station. In the municipal plan of 1997, construction of another 1200 dwellings in the district and location of commercial activity close to the metro stations has been suggested.

Table 6.1   Density of Copenhagen districts:
1: Inhabitants/ha.
2: Inhabitants and no. of employed/ha.

	Inner Østerbro		Outer Østerbro		Inner City		Inner Nørrebro		Outer Nørrebro		Vesterbro	
Surface incl. in calculation	1	2	1	2	1	2	1	2	1	2	1	2
Total Surface in ha.	72	118	70	102	55	274	178	249	185	230	88	152
ex. port areas	102	166	70	102	55	274	178	249	185	230	88	152
add. ex. railway areas	107	175	75	110	55	274	178	249	196	244	129	223
add. ex. roads	139	227	99	144	75	372	219	306	268	334	174	301
add. ex. green areas	203	331	139	202	92	461	272	381	296	369	189	326

The housing developments on the harbour front are primarily large expensive dwellings, whereas those dwellings on the former defence sites are smaller and have more reasonable price levels. In general, however, the housing built in the districts consist of large dwellings, which is in line with the municipality of Copenhagen's policy of attracting families and high income groups. In addition to this policy, the Municipality of Copenhagen has stopped building low income, "social" housing. Offices are primarily built on the harbour front, but also on a former industrial site near one of the local metro stations.

This current urban development has led to an increase in the district's population density as well as its physical density. Furthermore, is has led to an increase in jobs and a small increase in local employment. There was a small decrease in jobs in the beginning of the 1990s, but since 1994 more than 2000 additional jobs (equivalent to an 8% increase) have been created in Østerbro. Still, only about 15% of the residents are locally employed. There is an increase in the number of commuters out of the district, and especially out of the municipality.

Changes in the urban structure

This densification of Østerbro has consequences for the local urban environment. The high level of building activities has led to an increase in population density in the district by 5.9% which is higher than the increase for the municipality of Copenhagen generally. Part of this increase is due to an increase in household size in the existing households in the district. When one adds the 2000 jobs created in the district, the net human density of the district has increased further.

From a regional perspective, the redevelopment of these areas in the central districts makes sense. However, there is still urban sprawl in the region and thus some land loss. During the 1990s, the urban development tended to be drawn towards the outskirts of the region and at the same time towards the central districts of Copenhagen. Building activities in the suburbs and transitional zone have been limited. The housing built in the outskirts is mainly detached single family houses, whereas the housing built in the central districts are multi-storey apartments.

Another question with a more urban perspective is "In which urban district the densification should take place?" As shown in table 1, Østerbro has the same density as other central districts. All have a much higher density than the surrounding districts. Should these districts then become denser?

From a local perspective, the question is, "Which areas should be redeveloped, and with which densities?". This depends on the location in the urban structure. Several factors are important in relation to this question: the existing supply of housing, green areas, services, and location of these areas in the urban structure. In the case of the current densification of Østerbro, one could ask whether such questions have determined which areas were redeveloped. In fact, such questions have not been asked. The redeveloped areas are the ones where developers have come with a project. Location in the urban structure has never been a determining factor for development.

Figure 6.2 Offices and housing on the harbour front. The form and architecture of the housing on the right refer to the former warehouses

When we study the actual location of the new developments in the urban structure and the physical and functional structure, most of the developments are located in existing high density areas well served by public transport. The conflict of interests, in the function of the area has mostly been with recreational interests. In the area located near the metro station, local residents have fought for a park. This area has more than 1 km to the nearest park. From a regional perspective the location is perfect for high density development due to its proximity to a metro station in an existing high density area. Recreational interests, and the share of housing have been the central issues on the port areas. An overall assessment of the redevelopment of the district shows the displacements in the urban structure following these redevelopment projects (figure 6.3).

Figure 6.3   Displacements in the urban structure

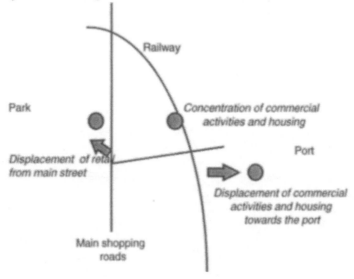

Transport

The redevelopment of the port areas resulted in a new road along the railway to connect the harbour redevelopment with the northern part of the district as well as with central Copenhagen. It was also the intention that this new road should relieve traffic on existing main roads.

During the last decade there has occurred an increase in the level of transport on the main roads of the district. The higher level of transport on several of the main roads has consequences for local air pollution, noise levels, and the physical and visual barriers that substantial traffic create. The new road along the railway to the port areas has reduced traffic on other roads but at the same time has created an additional physical barrier between the district and the port areas.

There has been an increase in traffic accidents during the period 1994 -1996 when the major redevelopment started. These accidents have taken place on the roads where there has been an increase in traffic, but also on the main shopping streets in general. So the correlation is not a simple one.

Why is this increase in transport in Østerbro occurring, when all theories on the compact city suggest that central localisation should lead to a lower transport level? If we study the volume of transport for some of the new housing developments in the district, we find that a central location of housing developments does not necessarily lead to a lower volume of transport. In three large new housing developments in the district, we find substantial differences in the volume of transport. These differences are probably related to the income level of the households, the location of workplaces, and the general lifestyle of the different households.

In general, traffic has increased in the municipality. This is due to several factors, including an increase in workplaces in the municipality. Furthermore, there has been an increase in car ownership in the Copenhagen districts. This is probably related to the fact that the Municipality of Copenhagen have succeeded in their policy of attracting families and high income groups. This stated policy has been effectuated through urban renewal, where flats where enlarged, and through building large flats in the new housing developments. Some of these high-income groups may not change their transport behaviour substantially when moving to central districts. Their workplace may be located far from public transport and their leisure activities probably do not change due to their new residence, if they have moved in from the suburbs. The increase in commuters out of the district indicates this. We thus have some explanation for the increase in transport in the district, and can conclude that densification of housing does not necessarily result in a more sustainable city, in terms of reducing transport. Lifestyle is an essential factor in environmental issues. From a regional point of view, the question is whether these groups would have a higher volume of transport if they lived in new housing developments in the urban fringe.

Green areas

The densification of the district has not caused a major loss of recreational areas. However, densification is not only a question of how dense an area becomes, but also a question of the loss of possibilities. On the former industrial site located next to one of the metro stations, redevelopment is taking place with an office building being built nearest the station and a 6 storey housing development on the rest of the site. The offices are completed, but no investors have so far been interested in the housing part of the project. Although the area has never been accessible to the public, the redevelopment project has encountered resistance in the local district. Local recreational interests in the district have fought for a new park, the area being poorly served with local recreational facilities. The distance to the nearest park is more than 1 km. Local residents overturned the fence running along the development area when the housing project was delayed. Subsequently, they bought plants and trees for the area. Development of this site is an example of the conflict between regional and local interests, and in the current urban development of Østerbro, local interests lose. For the residents, the redevelopment scheme is an example of lost possibilities for creating a local park. On the port sites, recreational interests are also having a hard time. Many of the proposed developments have a high proportion of offices and are designed as closed blocks around the port area, creating a wall against the housing district. Thus, the current urban development does not support the general aim of the local authorities to provide public access to the quays as part of the recreational policy.

Green areas in relation to the new housing developments are also hard to obtain. In relation to one of the projects, the local authorities had to sell a local municipal square to the developer, due to the fact that he was not able to fulfil the demands of the local plan proposing a maximum plot ratio of 150%. Thus, the residents lost the possibility to obtain more open space in the part of the development where the housing was located. In another development, the local authorities agreed to regard the development area as a single land register number. The result is that the open space of the area can be calculated as a whole. Consequently, the housing area has some of its open space 500 meters away! When the local authorities make agreements like this, the quality of the housing provided in the compact city decreases. And this may be a problem if we are to create quality examples of compact city developments.

Urban space and the use of space

The redevelopment and densification of the district have created new urban spaces and thus the opportunity for new uses of space. Quality open space is a crucial component in the vision of the compact city (Crookston et al., 1996). The question is whether the redevelopment has resulted in creation the of new good quality space in the district. One way to examine whether this is true, is to assess whether these new spaces have resulted in new recreational and cultural spaces for the inhabitants of the district and to study the use of space.

The redevelopment of the former military area includes development of a square and a shopping plaza. The local authorities sold the square to the developer as a part of their agreement, thus delegating responsibility for the redevelopment of the square to the developer. The local authorities had had some intentions with this square; it is centrally located in the district next to the main road, the local swimming pool and sports facilities. The local authorities thus saw the redevelopment of this square as an opportunity to create a key public space for the district.

When we study the shape and use of the square today, the authorities' intentions are not quite fulfilled. The square is rather an area one passes by in order to get in to the shopping plaza built by the developers and which lies behind this square. The signs at the square tell the pedestrian to go further. There is no reason to stay. There is a minimum of benches, designed as monuments rather than for sitting. The use of these benches is very low.

The use of the shopping plaza behind the square is higher. Here, there is a café, a large supermarket, small specialised shops and some cultural institutions. With one exception, the cultural institutions were present before the redevelopment. The shopping plaza is a sort of appendix to the traditional main shopping street of the district. It is created as a close where there is no common social surveillance. The buildings in this part of the area are only offices and shopping facilities, which could create an unsafe area in the evenings. More police could help, but it will not solve the problem, as Jane Jacobs already observed 35 years ago. If we are to create high quality urban space, more attention should be paid to this subject.

The local authorities have a crucial role to play in this situation. They might have stated some demands regarding the quality of the space created in this redevelopment, especially the area they had sold. But they did not. The reason for this lack of action might be found in their grounds for selling the square. The main reason the square was sold was that the developer, with his proposal for a local development plan for the area, discovered late in the process that he was unable to hold to the restriction on a plot ratio of 150%. If the local authorities sold him this square, the problem would be solved... In this case, focus has been on the interest of the developer rather than on the common interest. Andersson, in chapter 4 of this book, has noticed that a consequence of this process is a slackening of local government responsibility for the urban landscape. And who will then attend to the interests?

The new spaces along the port have been essential to the public and political discussion of the redevelopment of this area. Today the area, called Langelinie, includes a waterfront promenade, an exclusive shopping arcade, serving cruise passengers, and "public" access to the waterfront in general in the area. The flats in the area are quite exclusive, which in addition to the shopping arcade for cruise tourists leads to local political response such as; "Langelinie for whom?" and the creation of a foundation, "Langelinie for Copenhagen Inhabitants!" The area has been a recreational area for Copenhageners since the creation of the port. In relation to this redevelopment, the fear has been that the less well-off were squeezed out of the area.

Figure 6.4   The square today: More attention has been paid to what is behind the square, than to what could actually happen on the square. In the background: the redevelopment of the former military area with the entrance to the shopping plaza

One of the ambitions in the compact city vision, as stated by the EC in the Green Paper on the Urban Environment and in the UK Sustainable Development Strategy, is to support high quality local facilities by promoting more compact urban development. as we have seen, however, the urban space created in many of these Østerbro developments acts contrary to such objectives.

Social segregation

Some of the developments support continuing social segregation in the city. The urban spaces created are exclusive, and the flats built especially on the waterfront are expensive, and generally large flats. The policy of the local authorities has been to attract families and high-income groups back into the city. Among others, Smyth (1996) has pointed out that the compact city policy and urban gentrification in inner city districts may lead to social exclusion. Higher price levels in inner districts, a result of urban renewal and gentrification, have led to displacement of the disadvantaged out of the core and inner districts into the transitional zone. Smyth underlines the evidence of this trend, which has been observable since the mid1980s, when gentrification of inner city districts began to accelerate. In Copenhagen, the current urban development supports this trend, and the new housing developments which have led to densification, along with the stated housing policy, have certainly supported this trend. The question is whether this is a sustainable social development.

One could question whether the current trend results in social diversity, as some have claimed in light of the compact city. On the other hand, the dispersal of the urban area that we have witnessed in the past decades has increased gender, age and socio-economic differences. Thus, Nystrøm (1995) has concluded that dispersed urban development is not socially sustainable. The problem of current urban development rather, is that it continues to support trends in the socio-geographical pattern, but with the opposite geographical direction. In some areas this may lead to higher social diversity, but in others it will not. Again it depends on the scale on which one studies the transitions.

The social imbalance that the expensive housing could create has been a central issue in the objections of politicians and local residents to some of these developments. The social acceptance of these projects and the densification can be studied by examining the objections raised in the period of public hearings about the local plan. In relation to all the projects there have been several objections to the high density of the developments and also objections to the functioning of the area. In general, however, redevelopment of the areas has been agreed upon. The question, as mentioned before, is how high a density should be accepted. Perhaps the acceptance of densification is higher in already high density districts, as indicated by Finnish research (Lehtonen 1996).

Figure 6.5 The housing developments on the port areas have already created their own private space. Although it is possible for the public to walk along the port, the domain created and the signs that have been put there tells one to go away

Service

A central argument for some people in relation to the debate on the compact city has been that a more compact urban development supports local services. However, current development of services in Østerbro indicates that this is not always true. Supermarkets seem to be an inevitable part of densification projects in Copenhagen and in other Danish towns. One has probably to understand the character of these old urban districts in order to understand the problems with the supermarkets. Typically, the supermarket has a parking garage, something very unusual for retailing in the central districts. And they have a giant scale; 1000-2000 square meters and a wide choice of goods. In general, they have another character than the ordinary retailing in central districts, which can be described as smaller specialised shops and food markets. Problems may arise for the existing local retailing. In addition, it is more likely that residents use their car to come to this supermarket. On the traditional shopping streets, where it is difficult to park, local residents typically walk or take their bikes when shopping in these shops. Developers have a financial interest in incorporating large retailing units in their developments. In this case, several local politicians objected to this large supermarket in a central district. Nevertheless, a majority of the municipal council voted for the local plan.

In general, there exits a trend towards larger units in Danish retailing. This has been a fact during the last 20 years. In these central districts, however, the continuing developments of this kind create transport problems when combined with good parking and access by private car, not to mention development of housing districts without any kind of services.

Changes in retailing in the district may also change the balance of retail in the city. According to the local authorities, however, the problem of this particular district was that they did not shop in the local district, but rather in central Copenhagen. Attracting local customers to new local retail outlets would thus benefit the district. This may be true. But some of the retail trade in the district has a larger hinterland than the district. This is true of the exclusive shopping arcade on the waterfront, the large computer store by the metro station and maybe also for the large supermarket. An imbalance is thus created in the internal hierarchy of retail trading in the city as a whole, and contradicts the intentions of the compact city in creating high quality local services.

Another problem in relation to local services has been social services. The social services in the district have not been able to keep up with the increase in the need for child care, for instance. There has been an increase in families moving to the district, and an increase in households with children staying in the district instead of moving out. This increase has not been followed up by a corresponding increase in social services such as child care and schools, which has led to a substantial critique of the local authorities.

Historical heritage

The historical heritage has been a major issue in relation to the different redevelopment projects in the district. In relation to the four large redevelopments, loss of historical buildings is a fact. In two of the projects, buildings which were listed as protected buildings were demolished "by accident" by the developers.

Figure 6.6 In the redevelopment of the former military area, the entrance, walls and some of the buildings have been preserved. But one listed building was demolished by the developer over night. It did not fit into his project

In other cases, the local authorities agreed with the developer that the building could be demolished in order to be able to create a more "perfect" architectural concept. But in all the cases, developers have pressured local authorities in order to see their interests served. And with success. According to Larsson, the cultural heritage is still under pressure in cities and towns. Although there is a growing interest in sustaining the cultural heritage, modern urban development is still marked by the interests of property owners and developers toward maximising profits. And local authorities are most often interested in the localisation of workplaces and housing (Larsson, 1995). This is not only a question of preserving a few old buildings, but of preserving urban environments and the perception of the district - the spirit of the place. This has been an underlying discussion in some of the developments. The radical redevelopments change the spirit of the district from a district with harbour and partly industrial areas, to high class financial and housing areas. Copenhagen for whom?

Planning conflicts

This case study of the current densification of a Copenhagen district, indicates several problems in relation to the local urban environment, when applying a compact city policy. An essential cause of these problems may be found in the way the local authorities have handled the development projects, and less so in the compact city policy itself.

In studying the current urban restructuring of Østerbro, interviews with local authorities have revealed the pressure placed on them by developers when negotiating local plans. The result has been that the developers interests have been favoured and the public interests suppressed. We have seen this in connection with public areas, green areas and cultural heritage sites. Local residents have called attention to this fact in relation to the planning of these projects. But this has not altered the final result. These conflicts are not new, but in the current development of cities in a competitive era, the conflict seems to be more severe.

Planning in Copenhagen

The concentration of housing and commercial activities in the municipality of Copenhagen is occurring due to several circumstances. One can see the concentration in light of gentrification of inner city areas, or in light of the economic processes of change. However, from a planning point of view, one historical fact is especially interesting. Copenhagen has had a long-standing interest in building as many housing developments as possible with its municipal boundaries. This fact has become more significant after the resignation of the 1960s, 1970s and 1980s, when the municipality experienced a considerable decline in population and a corresponding loss in tax revenue. The current increase in population should also be seen in light of this decrease during the past decades. In Greater Copenhagen, there is an ongoing competition between the municipalities to attract taxpayers and commercial activities. The current concentration of activities in Copenhagen can be seen in this light. This goes hand in hand with interests in densification of the urban fabric due to environmental concerns, but we must question whether these concerns will be fulfilled if current developments continue.

In order to have a co-ordinated planning and transport policy, a co-ordinated planning action and administrative organisation in the urban regions is necessary. During the restructuring of the Danish municipalities in 1970s, the main argument used to define the municipalities, was "one town - one municipality". Except for Copenhagen. The greater Copenhagen area still consists of 50 municipalities and 5 regional authorities (the central municipality of Copenhagen has a dual function as local and regional authority, as does the neighbouring municipality of Frederiksberg). A Greater Copenhagen Council has co-ordinated the planning of the entire region, but this council was abolished by the conservative government in the late 1980s. Today the co-ordination of planning in the region is carried out by the state in co-operation with the five regional authorities. The question is whether the current organisation of the region is sufficient in order to have a co-ordinated planning. The current urban development, with both urban sprawl and urban densification, do not provide sustainable urban development. Nor do these trends accord with the intentions of the National Planning Guidance.

Conclusions

Several lessons can be learned from this case study of current densification of a Copenhagen district. The conflict between regional and local environmental interests has been discussed in relation to the compact city, but competition, organisation, and the conflicts of implementing this strategy have rarely been discussed thoroughly. This case study has shown that the critique of the compact city concept has had some legitimacy. There are many good intentions, but they show little success on the part of the local authorities with the current situation in Denmark. The local authorities do not work with the compact city strategy as their overall planning strategy. And the case study has shown the kind of conflicts that arise when implementing a compact city strategy without having a co-ordinated planning approach, comprising housing, transport, environmental and social issues. Furthermore, the case study has shown the conflicts between urban densification as it is now occurring, local environmental and social concerns and the interests of developers. Anderson (1998) has described this phenomenon as "the illusion of urban renewal as an integration in which the basic attribute of the urban space is a richness and variety of relationships between people from different social and consumption groups falls down because of solutions dictated by the power of capital". If the compact city is to provide a more sustainable urban development, local and regional authorities must balance the strong financial interests in current urban development. And as densification is now occurring, this is not the case. In relation to densification in England, Breheny (1992) has pointed out that "the town-cramming that we have witnessed is the result of a piece-meal, largely unplanned activity, carried out without regard for environmental consequences." We can conclude that if urban den-sification is to bring more sustainable development to our cities, a more thoroughly planned densification process is an urgent neces-sity.

Back to the city?
Trends in inner city population in some European cities.

It seems as if we can trace a "back to the inner city" trend in several European cities. London has experienced a rise in population during 1983 to 1993 period which was fastest in central London, where the total resident population rose by 4.9 percent compared to 1.0 percent in Outer London. In some of the central London districts, population has grown by 18.4 percent. Some of this growth reflects the redevelopment of areas such as the Docklands. Between 1983 and 1993, the age groups which most increased their share of the population in Inner London were those aged 25 to 34, but the proportion of children also increased.

Stockholm has also experienced a rise in its inner city population during the 1990-95 period, where the resident population grew by 8 percent, compared to 4 percent in the Outer Stockholm area. Inner and Outer Stockholm have had nearly the same increase in the total stock of dwellings: 2.5 percent. Thus, Inner Stockholm has had a more rapid increase in the size of households than Outer Stockholm, which could imply that Stockholm is following the same trajectory as London.

Oslo has seen a rise in population in all its inner city districts during the period 1989 to 1998. In the inner city districts, the total resident population rose by 12.1 percent, compared to 9.7 percent in the Outer Oslo districts. In 1997, 40 percent of all new dwellings were built in the inner city districts, and more than 50 percent of all dwellings built in Oslo were flats. This indicates that building activity has been dominated by urban rather than suburban development.

The rise in inner city populations reflects several trends. In Copenhagen, inner city regeneration and urban renewal certainly explain some of the increase in population. The larger flats and better local recreational areas have helped keep young families in the city. London statistics indicate the same trend. In cities such as Oslo and Amsterdam, a compact city policy in relation to inner city regeneration has certainly had a positive effect on the rise in resident population. The growth in dwellings in inner city districts will inevitably create a rise in resident population.

Gertrud Jorgensen and Anne Skovbro

References

BREHENY, M. (1992) The contradictions of the compact city, a review, in M. BREHENY (Ed.) *Sustainable Development and Urban Form*, pp.138-159. London: Pion Limited.

BREHENY, M. (1996) Centrists, decentrits and compromisers: views on the future of urban form, in M. JENKS et al. (Eds.) *The Compact City - A Sustainable Urban Form*, pp.13-35. London: E. & F.N. Spon.

BURTON, E. et al. (1996) The compact city and urban sustainability; conflicts and complexities, in M. JENKS et al. (Eds.) *The Compact City - A Sustainable Urban Form*, pp. 231-247. London: E. & F.N. Spon.

CEC (1990) *Green Paper on the Urban Environment*. Brussels: Commission of the European Communities.

CROOKSTON, M. et al. (1996) The compact city and the quality of life, in M. JENKS et al. (Eds.) *The Compact City - A Sustainable Urban Form*, pp.134-142. London: E. & F.N. Spon.

FOUCHIER, V. (1998) Le densités urbaines et la mobilité, in UN/ECE-HBP and Ministerio de Fomento (Ed.) *Proceedings of the Eighth Conference on Urban and Regional Research*. Madrid: Ministerio de Fomento.

GRAHN, P. (1991) Om parkers betydelse, *Stad & Land*, 93, Alnarp: Movium/Inst. för Landskapsplanering, Sveriges Lantbruksuniversitet.

GUTTU, J. et al. (1997) *Boligfortetting i Oslo, Konsekvenser for Grønt struktur, bokvaliteter og arkitektur*. Oslo: NIBR.

JACOBS, J. (1961) *The Death and Life of Great American Cities*. New York: Random House.

LARSSON, B. (1995) Cultural heritage, local identity and commercial interests in comtemporary city centre planning in Sweden, in Ministry of Energy and the Environment *The European City - Sustaining Urban Quality*. Copenhagen: Ministry of Environment and Energy.

LYNCH, K. (1981) *Good City Form*. Cambridge, Mass: MIT Press.

NAESS,P. and ENGESAETER, P. (1992) *Tenke det, ønske det, ville det - men gjøre det?* Oslo: NIBR.

NAESS, P. (1994) *Energibruk i 22 nordiske byer*. Oslo: NIBR.

NAESS, P. et al. (1996) *Bærekraftig byutvikling*. Oslo: NIBR.

NEWMAN, P. and Kenworthy, J. (1989) *Cities and Automobile Dependency. An International Sourcebook*. Victoria: Gower Publishing Group.

NYHUUS, S. and THORÉN, A.K. (1997) Den grønne og tette byen, *Plan* 1/2. Oslo.

NYSTROM, K. (1995) The diversity of the urban environment - as a result of planning or of laissez-faire? in Ministry of Energy and the Environment, *The European City - Sustaining Urban Quality*. Copenhagen: Ministry of Environment and Energy.

ORRSKOG, L. and SNICKARS, F. (1992) On the Sustainability of Urban and Regional Structures, in BREHENY, M. (Ed.) *Sustainable Development and Urban Form*. London: Pion Limited.

OWENS, S. (1994) Can land use planning produce the ecological city? Town & Country Planning, June, London.

OWENS, S. (1992) Energy, Environmental Sustainability and Land Use Planning, in BREHENY (Ed.) *Sustainable Development and Urban Form*, pp. 79-105. London: Pion Limited.

RAEDBERG, J. (1995) Termite's heap or rural villages? The problem of urban density and sustainability, in Ministry of Energy and the Environment *The European City - Sustaining Urban Quality*, Copenhagen: Ministry of Environment and Energy.

SMYTH, H. (1996) Running the Gauntlet - a compact city within a doughnut of decay, in Jenks, M. et al. (Ed.) *The Compact City - A Sustainable Urban Form*, pp. 101-113. London: E & FN Spon.

THOMAS, L. and COUSINS, W. (1996) The Compact City: A successful, desirable and achievable urban form? in Jenks, M. et al. (Ed.) *The Compact City - A Sustainable Urban Form*, pp. 53-65, London: E & FN Spon.

7 Public Places and Urbanness

DOMINIQUE JOYE AND ANNE COMPAGNON

Introduction: Public places and urbanness[1]

Speaking of public places within the city raises a host of complex issues. First, there is the wealth of meanings ascribed to the notion of "public place". Town planners and social scientists approach the concept in very different ways: one discipline studies the physical development, while the other regards public space as a political construct. It is our conviction that both approaches are necessary to gain an understanding of the urban arena. This link between the social and morphological features is one of the basic themes of this chapter.

The approach to the political debate surrounding public places is once again being challenged by the ever-topical issue of citizen participation in the management of life at the local level. It is true that local participation has always been important. It is also true that urban movements have been included in the "new social movements". However, with the changes occurring in states and on the political scene – not least of all in the role of parties – local participation has been significantly transformed (Hamel, 1998). On one hand, the definition of the main political challenge has evolved from "politic" to "policies" (Leca, 1997), from a broad ideological debate to sectorial discussion linked to a particular stake. One of example of this is the appearance of local referendums in Italy, France and Netherlands, to name just a few. Furthermore, the aspect of participation is not only political, but also social, since integration implies the capacity to act on one's environment. For this reason it is essential to incorporate the political dimensions into social integration, a task which constitutes one of the major

---

[1]   Translation by Catherine Smith.

123

challenges to the research being carried out on citizenship in the Anglo-Saxon sense of the word (Adriaansens, 1994).

The approach to public places as physical spaces is closely linked to the elements which enable people to live together. As mentioned by Lofland (1993), Louis Wirth stated early on that «the juxtaposition of divergent personalities and modes of life tends to produce a relativistic perspective and a sense of toleration of differences.» This question of juxtaposition is important today as numerous authors refer to such notions as "community", "social capital" or trust in order to evoke the idea that public places have the potential to "improve" the quality of life in a city. Today still, increasing importance is attached to local integration in the context of building a community. Hill (1994, p. 3) for example writes:

> "The urban arena is used in this book as the setting for citizen involvement, and for the exploration of the meaning of community. Community may be based on the shared experience of place, on attitudes and loyalties, or on common interests. Individuals are members of a number of communities in these different senses, all potential avenues for political participation. Community, however, is a contested notion... Whatever the perspective, some concepts of the community as the shared experience of place provides the justification of the locality as the arena for the exercise of citizenship."

While this may be true in certain cases, or to a certain extent, the conditions in which this may be true must also be closely examined. In short, this hypothesis should be studied empirically.

This type of subject specifically refers to the way in which different groups live together in the urban environment, that is, the notion of urbanness. According to Françoise Choay, "urbanness is the relationship between, on the one hand, a built-up place and its spatial environment and on the other, the ability of the group living in it to generate social and sociable relationships". Though the notion of "urbanness" involves a complex combination of elements, its main requirements are a minimum threshold of density and diversity". Let us note that Lofland also raises a similar question: "I want to inquire into this matter of how, if cities manage to create civility in the face of heterogeneity, do they do it? To paraphrase Donald Olsen, what do cities "do" to people to make them urbane?" (p. 94) The idea is there that public places are an important indicator of city life, in both its urban and its social aspects. How are these places a reflection of urban identity? Just how well do they represent the cohabitation of various groups? To what extent are they forums of expression? And finally, do they provide a means of countering the forces of fragmentation which threaten contemporary urban environments?

In this chapter, we will first take a closer look at the role of public places in cities, after which we will empirically examine how the use and the forms of these spaces are related to the appreciation of cities today, as well as how they may possibly create a new type of urbanness.

The role of public places: theoretical considerations

The context of metropolisation, characterised by increased daily mobility and by a change in the global as well as the urban focus is underlying our research on publics places. Indeed, many studies on globalisation have shown that there are two scales involved in the metropolitan processes: one is an attempt on the part of cities to position themselves with respect to a world market, the other is a rearrangement of spaces internally, with social as well as spatial consequences. This is where public places can prove to be pertinent measures because they encompass social as well as spatial aspects; they incorporate the way in which a city is seen and the practices of its dwellers, and they are at the cross-roads of the local and the global.

In more specific terms, public places are part of the image of a city. This hypothesis has at least two sides: for the residents, as an extension of their living space, public places are linked to both their daily practices and to the image of the neighbourhood and its surroundings. One has only to look at the importance attributed to green spaces in studies carried out by local authorities on the satisfaction of residents (Gachter, 1998). For the authorities, such places represent an opportunity to present a favourable image of the city from a point of view of urban marketing (Biarez, 1998, p. 72). In this way, public places also serve to attract visitors to the city. It is unnecessary to highlight that this form of reflection is also a function of the "spirit of the times". A priori, a certain number of cities (including Lausanne, Switzerland, for which we have data available) have progressed from prestigious, highly visible developments in the 1980s to more modest developments located in different areas of the city. This does not conflict with the idea of "beautifying the city" but is undoubtedly a sign of a government which is as attentive to residents as it is to public founds.

Furthermore, according to certain authors (McKay, 1996), the local elite networks, their location in the city centre in continental

Europe, and the dominant values also influence the choices for redefining central spaces. Finally, these spaces are part of the perception of the neighbourhood, which is not a negligible factor. Stigmatisation of a particular neighbourhood contributes to the feelings of exclusion which may be experienced by its residents (Joye, Huissoud, and Schuler, 1995).

Public places are also the point of convergence of the domain of planning and the daily lives of individuals, and as such, constitute building-blocks of participation. It is not by chance that the aspects of town planning alone form an important part of local political life, if one judges by their occurrence in the popular votes so frequent in Swiss cities.

Most importantly however, public places are at the heart of a series of heated debates on the current functioning of contemporary societies. We would like to concentrate on four aspects that are closely related to one another:

First of all, there is the idea that social interaction is desirable. The reverse side is that exclusion and marginalisation are the specific results of severed ties and the breakdown of social interaction. It is for this reason that the issue of social capital is so prominent today. The underlying theory is that participation in a multitude of networks facilitates both the integration of the individual and the smooth functioning of the community. Putnam's idea of social capital therefore refers to a micro-sociological as well as a macro-sociological dimension, and postulates that societies with a plurality of networks function more efficiently. Furthermore, it would appear that it is social capital that promotes interpersonal trust, and not the other way around (Brehm and Rahn, 1997). Similarly, withdrawing into private circles, or "cocooning", weakens the social link (Remy and Voye, 1981), (Sennett, 1982). The use of public spaces is believed to promote integration. The real issue however, undoubtedly goes far beyond the narrowest sense of the concept of public places. The title of Dominique Schnapper's last book illustrates this: *The Relationship to Others at the Heart of Sociological Thought* (Schnapper, 1998), underscoring the importance of the link between individuals in sociological thinking.

In parallel, the idea of belonging to a community is also a topical issue. Some argue that a society which returns to the notion of "We", as opposed to the "I" proposed by liberalism, functions well (Etzioni, 1995, p. 18, see also Bellah, 1997). Putnam also views the community in a very favourable light, and Levi even uses the word "romanticised" to describe the Putnam's perspective in this context (Levi, 1996, p. 51). This debate brings to the fore the old argument of what should be "public" and what should be "private"

(Kaminski. 1991). In the context of the city, this debate cannot be separated from the discussion of social segregation, whether it is a question of the privileged classes moving into "good residential areas" or gated communities – the side of the coin most often quoted – or, rather, the issue of its negative counterpart involving the image of the ghetto. Researchers traditionally tend to represent segregation as negative, and integration as positive. This is true not only for the social sciences, but also for the urban managers who for so many years have favoured the mixture of jobs and housing, partially with the aim of reducing travel time; this proved not to work in the majority of cases, since residential localisation strategies are often separated from issues concerning jobs, especially when several working individuals are comprised in a single household (Gaudin, Genestier and Riou, 1995). It should nevertheless be noted that "rubbing shoulders" with the other person does not necessarily produce a sentiment of empathy or sharing. Contact others, by making differences tangible, can also lead to a situation of exacerbated prejudice and potential conflict. Integration can take place in segregated spaces as much as in mixed spaces, although at different levels. In short, integration and appropriation can be positive, as long as the different groups of residents can agree, implicitly or explicitly, on a certain number of rules to be respected. Let us recall here that Putnam's "social capital" includes the fact that the other groups "follow the rules". Similarly, according to Goffman, the sharing of a space involves a feeling of trust towards the other person.

The idea of subsidiarity with regard to the State, which is of course a political dimension, is also present. To a certain extent the sense of community, and perhaps also the sense of association (Hirst, 1998), originate in the idea of the decline or the failure of a certain form of State. In other words, the idea also exists that the State alone is not able to maintain or to limit social fracture, but that it has a subsidiary role. There is a need for solidarity, stemming from proximity or proximity networks. But it is unnecessary to point to similarities between the issue of the end of the Welfare State and the one of regulation, especially with respect to government and the association of both public and private partners (Biarez, 1998). And yet public places specifically raise this issue of proximity.

Thus, from a theoretical point of view, several approaches provide grounds for the hypothesis that the development of public places promotes the smooth functioning of the city because they allow for interaction, exchange, mixing with people who are different, and the functioning of proximity networks. Nevertheless, a certain number of precautions must be taken.

First, the issue of mobility. The majority of inhabitants no longer remain confined to a given space, but their habits with respect to mobility lead them to cover a vast portion of the city's territory, starting with the central areas. The question of accessibility, but also of the means of accessibility, whether by «public» or «private» transportation, becomes crucial. This further raises the issue of co-ordination between public places of proximity and central public places, or the question of co-ordination between networks and territories (Offner and Pumain, 1996).

Social position in the broad sense is also a determining factor, whether for accessibility or for the resources required for the different types of participation. In this context, Putnam has been criticised for having a normative approach which coincides with the dominant lifestyle of the American middle classes (Pollit, 1998). More generally, when one considers that public places are neither visited nor regarded in the same manner by different social classes, the issue broadens to encompass the management of social fragmentation and, as a consequence, that of solidarity. Notions of solidarity, trust, identity, and cohesion can therefore be examined only in relation to participation in the socio-political system; however, this participation is linked to social position.

Daily use but also representation of neighbourhood is of importance. The relation to the neighbourhood and to the built-up environment is not only a function of localisation, but also of relocalisation defined as the value attributed to the local space, and it involves not only past experience, but also knowledge of the environment (Lamarche, 1986), (Joye, Huissoud and Schuler, 1995). As a result, the use of central places is inseparable from the value judgement of a city.

In summary, all these remarks raise the issue of solidarity and fragmentation in modern cities, and it is no coincidence that two recent books bear nearly identical titles which refer to urban fragmentation (May et al, 1998; Haumont and Levy, 1998). However, the following question has been asked (May, Sector and Veltz, 1998, p.22), "Are cities truly experiencing fragmentation? Does the expression, by its very formulation, rather not boast of the spatial dimensions of a global trend, transversal to our society, whose spatialised symptoms are the simple expression of the new urban character of our societies? Should we now begin to refer to 'fragmented cities' or 'fragmented societies'? The question is not a simple one, for it is obvious that there would be no point in trying to reduce, by means of strictly spatialised policies ('physical development', for example), processes which are the result of more expansive logics. This could construe ambiguity in so-called 'urban' policies, by attempting to have them solve problems outside of their jurisdiction." In other words, it is futile to speak of local development without taking into consideration the social context, just as it is not sufficient to develop a public place in the architectural sense in order for the magic to work and produce social relations. This viewpoint highlights the need to maintain socio-political and spatial components in a global plan.

Analytical diagram and hypotheses

On this basis, we can draw up an analytical diagram which shows the relation between the main elements to be examined, and which will highlight a group of hypotheses to be examined.

Analytical diagram

A number of comments and explanations shall be given for this diagram (Figure 1). It includes concepts related to social structure, attitudes, values, practices, and perceptions. It can be read in at least two ways: first, the use of public places is located in the centre, and the various explanatory elements converge around this centre. Secondly, the graph can be read from left to right, with the social position and the various uses and perceptions of the city illustrating political participation and social integration. Furthermore, the explanatory aspects are broken down into the political components at the top, and the geographical and morphological considerations at the bottom; the purely sociological variables are intermediate in this system.

Let us take a quick look at the main aspects of the diagram.

Taking the sociological side first, it is clear that strictly speaking, social position has to be measured using the classical variables: gender, nationality, lifestyle and socio-professional categories (Joye and Schuler, 1995) which refer directly to education and employment, i.e., in the words of Bourdieu, to cultural capital and financial capital. Today, numerous studies of urban sociology are focusing more on issues of exclusion than on those related to the position within the social layers themselves. We nevertheless support the thesis that these two ways of examining inequalities are complementary, and must accommodate one another (Levy et al., 1997).

Lifestyle is an essential element in the relation to others and to space. Here again, there are two viewpoints: firstly, lifestyles characterised by active free time, as opposed to those lifestyles where free time is spent inactively, and second, lifestyles encompassing only the household, as opposed to those mainly involving public places.

Spatial and socio-political fragmentation are measured via feelings of inequality, antagonistic groups, and differences in accessibility to policies. We have focused here on this concept rather than on its opposite, cohesion, whose content is often normative. Indeed, cohesion is often perceived as something under threat which requires reconstruction (Plan and Delevoye, 1997). The relationship between cohesion and social conflict nevertheless remains ambiguous, insofar as cohesion tends to deny conflict. We support the inevitable and even constructive nature of conflict, as long as a certain number of principles continue to be agreed on. Then the notion of social transaction (Blanc, 1992) can be regarded as a means of regulation rather than as a hypothetical cohesion which could have a tendency to *de facto* reinforce the position of a particular social group. The same argument has been formulated in the local development field (Joye et al., 1990).

The socio-political aspect is as much part of the relation to participation as the relation to solidarity. In this context, a few simple indicators must be developed to situate the actors with respect to values (right-left, post-materialistic).

The relationship with the political system may not appear to be of great importance to our problem. What is crucial however, is that the ability to act and control over one's environment shape the position which individuals adopt with respect to other people. This is all the more significant if considered alongside the fact that local development, including development of public places, constitutes one of the foundations of local participation. The ability to control one's environment and to have a say relates to socio-political as well as geographical concerns. This leads to the second major postulate of this paper: quality of life in the city and the ability of groups to share a space depends first and foremost on the ability to express oneself and to take an active stance. Again, this does not mean an absence of conflicts, but that there is perhaps agreement as to a certain number of norms pertaining to the expression and use of the space. At the political level, participation, which also represents a form of integration, governs the relationship with authorities and the ability to act in one's own interest, including by means of community action. This means that notions of solidarity and of trust can be dealt with. Trust in particular, whether in relation to authorities or to the other groups sharing the urban space, cannot be left aside, as it is considered by certain authors to be the permanent glue holding the urban structure together, and is closely linked to the social capital of which we spoke.

Figure 7.1 Analytical diagram

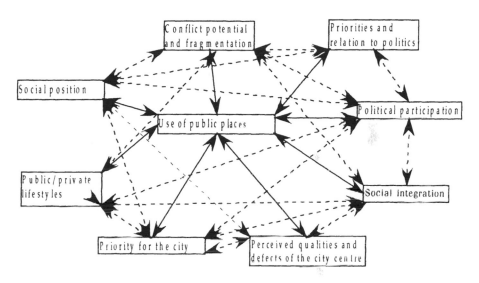

Social integration is an important intermediate variable, as yet another factor defined by social aspects and political participation. From a sociological point of view, trust is measured via networks of mutual knowledge, but it is also linked to the degree of attachment to the city and neighbourhood.

From a geographical point of view, spatial integration is a function of the view held by residents regarding their neighbourhood and their city. In other words, as studies on insecurity demonstrate (Roche, 1998), the way in which the built-up environment is perceived is a determining factor. We will approach the concept of spatial integration in two ways, one dealing with the city as a whole, and the other with the characteristics (qualities and defects) of the central places.

The principal variables remain the use and perception of public places. Here again, the issue involves two aspects. From a geographical point of view, how are they used? How does their geography shape their characteristics? Are these characteristics central elements of the city? From a sociological point of view, are they places of exchange and meeting? What is their symbolic value? Are they considered "key places" in life and the city? Are they worth defending? How?

In order to approach these facets of public places, a series of questions in this survey was devoted to the centre of the city, including the definition of its limits. We shall deal with public places in the centre of town, since their function is not limited to that of the neighbourhood, but to the entire city. Let us recall that since the 1980s, a time when there was talk of the veritable abandonment of the centres of certain cities, the attention of authorities was drawn to these centres as spaces which could prevent the exodus of residents to outlying areas, as well as focal points of local identity and a means of enhancing the cities' attraction with respect to competition. In short, the city centre was seen as a means of making urban life desirable (O'Connor, 1997). In addition, another series of questions focused on specific places in the city of Geneva with different characteristics. We will come back to this aspect.

Hypotheses

As we have mentioned, from this point, two different theoretical approaches can be put forward. The first proposes that an individual's perceptions and practices of public places are influenced by his or her social position, relationship to space, and by the image that the individual creates. The second approach reverses the relationship and considers the uses and images of public places as formative elements of the scenario of living together within the metropolis. Within this theory, acting on the development or the image of public places has the effect of acting on urbanness. In the first instance, we will refer to "defined space", and in the second, to "determinant spaces".

Defined spaces

In the first model, use of public places is determined by social position and by the perception of the city, but also by the relationship to other people and to public life. On the basis of this approach, a certain number of hypothesis can be proposed. It is indeed probable that a high social standing entails an outward-looking lifestyle, and therefore a more intensive use of the most central public places. Furthermore, a privileged social position is related to control of the socio-political environment. For this reason, these social categories can be expected to appropriate certain public places. In addition, their appreciation of the aesthetic may promote support for the construction by authorities of prestigious public places for the city. At the same time, the tendency to section off the privileged social classes into "nice residential areas" may lead to a selective use of public places, which in turn may deprive the image of the city of its dominant urban characteristics: the more privileged one is, the more likely "the Other" is to represent a danger. These two hypotheses are partially contradictory; however, two elements can explain their coexistence. First, it can be assumed that for some people, the image of public places is more important than the use which is made of them. For many, the primary purpose of a public place is to provide an opportunity for interaction with others, even if the principle of "reciprocal indifference" means that actual contact is rarely sought (Lofland, 1998). Secondly, reducing the issue to privileged social classes can be somewhat short-sighted, given the fact that people's behaviour varies greatly depending on whether their basic resources are financial or cultural.

Independently of social position, the use of public places is also related to lifestyle and the perception of the urban environment: the more "outward-looking" the lifestyle, the more the city as a whole will be appreciated, and the more intensive the use that will be made of public places. Similarly, the degree of integration in society and involvement in politics is directly proportionate to the appreciation an individual has of public places. Nevertheless, it should be noted that the appreciation of a public place is not systematically synonymous with more intensive use of that place; lifestyles and social position have to be taken into account.

Determinant spaces

What conditions are required for public places to contribute to urbanness? How can they strengthen social links and increase the performance of the city? Public places can be forums for integration as well as factors of exclusion. They can improve the image of a city and they can tarnish it. The following hypotheses illustrate this.

As a vector of the city's image to the rest of the world, a public place is sometimes used as an urban marketing tool to attract investment. However, this approach may lower the quality of life for residents, whose concerns are more locally focused. It can therefore be supposed that the type of public place selected for investment is related to the general attitude towards the type of development desired for a city.

In allowing the users of the city to mix with one another, even anonymously, public places forge social links. In this context, it can be affirmed that the qualities of a city are based on those services which allow exchange to take place without coming into conflict with the intensive use of a public place. A refusal to communicate or conflicts of appropriation, on the other hand, can undermine this role of a public place.

Public places, as forums of collective memory and local identity, can function as elements of social bonding. Nevertheless, contemporary lifestyles and modal practice can produce very diverse scales of reference. It can be supposed that very localised integration within a neighbourhood does not lead to either a favourable perception of the city or to a positive image of other social groups.

Central public places have the ability to make the city centre attractive. As such, they contribute to the cohesion of the whole and can add to the image of the compact city. For some people however, the morphological transformation of the city means that their experience of said city is restricted to the neighbourhood, and even to the outskirts and supermarkets; the city centre is excluded from their domain.

As spaces of community life, public places may be the objects of widespread perceptions and norms (as to their legitimate purpose and image) which facilitates their utilisation. This may be true for a specific place or for public places in general. But this consensus does not necessarily exist; it may be inhibited by the coexistence of diverse populations, since each of these has different (or no) references. These differences may lead to conflict and feelings of insecurity.

In the two models presented here ("defined spaces" and "determinant spaces"), public spaces appear as powerful social indicators, in that they provide a canvas for studying the relationships between people living in a city. However, it should be noted that the space experienced by an individual (i.e., the use made of this space) is not sufficient to evaluate the relationship with places, and may even distort the vision of this relationship. It is therefore necessary to examine the knowledge of an individual (i.e., the image of public places). We must further recall that social life encompasses more than what occurs in or around public places, and that social links may take forms other than those revealed in these «high places». Once again, this points to the necessity of including all the factors which influence community living.

Methodology

The study which will allow us to verify these models (at least to a certain point) was carried out in the most densely populated areas of the city of Geneva. Although Geneva is demographically relatively small, its area of influence is expansive, and an estimate of its inhabitants places the population at 600,000, covering two cantons and extending beyond the national borders. Geneva's cultural life and its social make-up are heavily influenced by the presence of numerous international organisations and head offices of large companies. This explains why empirical research has depicted the city as being of greater significance than its relative demographic size (Cattan et al., 1994). We should note, furthermore, that Switzerland has experienced true unemployment only since this past decade, which paints a social portrait that contrasts with that of other European countries.

Data were collected during a telephone survey carried out by a Geneva survey institute in the Autumn of 1998. The questionnaire comprised approximately sixty questions, the majority of which required a "yes" or "no" answer. The sample comprised 900 individuals selected at random from telephone lists, but controlled by quotas (neighbourhoods, age, sex).

Results and discussion

The results which we will examine here can be divided into three categories: the first two constitute a descriptive portrait of the importance of the public places located in the city centre and then those in three selected areas. Following this, we will be able to examine all the data selected to position the hypotheses of «defined spaces» and «link-spaces».

The city centre

From an empirical point of view, the city centre and its public places are highly desirable: 84% of the respondents found it either very or quite pleasant to spend time in the city centre. Two-thirds considered that public places were very important for the quality of life, while only 2% believed that they were of little or no import.

This view is widely held, and in fact, only those respondents 70 years of age and older found spending time in the city centre not to be pleasant; the opinion with respect to public places nevertheless remained unanimous. Similarly, there was no difference in opinion related to gender, socio-professional position, or nationality.

The centre of town is used by three-quarters of respondents for shopping and by two-thirds for leisure activities. In the case of leisure, the differences were more pronounced: social position had a slight influence, with managers and other white-collar workers using the city more often. Age also has a small influence: use decreased from adolescence to adulthood, after which it increased slightly during retirement, and finally decreased from age 70 onwards. It should be emphasised that this result is above all true for the city centre, which appears as a Mecca beyond criticism. The centre to a certain extent represents the city in its global dimension. On the other hand, specific public places within the city present more variations. We will deal with this point in another paragraph.

The question is: does the use of the city centre have an impact on the image of the city or of urbanness? Some factors would appear to confirm that it does. First, it is not surprising that when the city centre is used intensively, public places are perceived as being part of the quality of life. However these same places are also linked to the image of the city's future. The development models available to respondents were Geneva as a compact city (construction within the city), an extended city (construction in the countryside), neither model, or both models, although the last two choices were not indicated explicitly. Respondents also had the option of disregarding the question. A small majority (50%) opted for the compact city, 22.4% chose the expanded city, and 14.4% chose neither of the above solutions. It should be noted here that the use of the city centre produces a slight bias in favour of the compact option.

The city of Geneva is generally considered safe, but one-third of our respondents consciously avoided, after 10 p.m., certain streets or squares in towns they considered unsafe. Here again, the more the city centre was used, the less the respondent felt unsafe. In other words, this is an example of the well-known fact that insecurity exists as much as an image in people's minds as it does in reality.

This late demonstrates that there is a correlation between city living and the use of the city centre. There is even an effect on the expression of local solidarity, which is less developed when the centre is used less frequently.[2] This result, apparently paradoxical,

---

2   One of the survey questions was formulated as follows: when you think of the people in your neighbourhood, do you feel a great deal of solidarity towards them, a relative degree of solidarity, a small degree of solidarity, or no solidarity at all? The following replies were obtained: 14.4% felt a great deal of solidarity, 40% a relative degree, 26.8% a small degree, and 16.4% no

can be explained by the opposing situations of those who find themselves in a local space, close to their homes and private lives, and those who consider themselves to be in a metropolitan, or public, space. Use of the city centre apparently has no influence on the perception of authorities, nor on the appreciation of other groups in public places.

These remarks demonstrate that there is a strong correlation between the use of public places in the city and city living. They also show that these places represent a common image shared by all social classes; nevertheless, they leave two questions unresolved: is what we have outlined here true of all public places? What role does this use of public places play within the ensemble of the social characteristics and images which comprise urbanness?

Three public places

We refined the analysis by concentrating on three public places, in which we carried out detailed observations and surveys of the users *in situ*. The three places are described briefly below.

Place Neuve: the area of this square is defined by the surrounding buildings: the opera, the conservatory, and the art museum. These buildings date back to the nineteenth century, and the location is a prestigious and central one and important to the city's heritage. It lies close to remnants of the old city walls, the monument of the Reformers, and buildings belonging to the university.

Plaine de Plainpalais: located close to Place Neuve, this is a vast, partly green expanse. It is distinguished by the use made of this location and the main roads surrounding it. Plaine de Plainpalais is used for many diverse activities, which change according to the day of the week and the season; and is essentially the scene of popular gatherings rather than solitary activities. It is the site of the fruit and vegetable market, the flea market, exhibitions, circuses, etc.

Place des Volontaires: located on the banks of the river Rhone in a former industrial area, is the smallest of the three spaces. Today, one of the two industrial buildings surrounding the space is used as an alternative cultural centre, while the other has been converted into a concert hall used notably by the opera. Accessibility to this place has been facilitated by the construction of a passenger bridge over a new dam.

---

solidarity. The first two categories comprise a slight majority of those who use the city most often, and they comprise 60% of those who visit this area the least frequently.

Figure 7.2 Place Neuve and parc des Bastions

Utilisation of these public places is more varied than that of the city centre, both in terms of use and social profile. The first two locations have many similarities, since they are used relatively often by three-quarters of the respondents. Nevertheless, there exists a subdivision of social categories exists; less than half the workers said that they regularly visit the Place Neuve: this usage is globally intensive. There is also a more elitist profile in Place Neuve, and more workers apparently use the Plaine de Plainpalais. In both cases, more men than women use these public places, and this is all the more true for foreigners and for the under-thirty age group.

These differences would appear to be much smaller for the Place des Volontaires, partly because less people use this public place: only one-quarter of respondents admitted to using it on a regular basis. Those with greater cultural capital (students, white-collar workers, managers) are heavily represented here, and there are very great differences among the various age groups. Gender and nationality, however, show hardly any variations.

Figure 7.3 Plainpalais

How are these three places perceived? First, let us note that the Place Neuve was the most monumental and the most widely appreciated. The Plaine de Plainpalais was less appreciated than Place Neuve, and the Place des Volontaires, even less so. In fact, the appreciation of a place is directly linked to its use; if respondents used a place, it tended to be more well perceived. This result becomes even more apparent if social diversity is taken into account: indeed, if the place is used by a respondent, whether or not it is used by all groups or a particular group, it is appreciated by this respondent. On the other hand, if the place is perceived by the respondent as being appropriated by other specific groups, he or she tends not to appreciate said place if he or she does not use it. In other words, the use of public places would have us assume that users are more open to others, or that it promotes urbanness. Nevertheless, this result is placed in a more global perspective if the social position and the image of what constitutes the urban environment today are taken into consideration.

Figure 7.4 Place des Volontaires

From use to insertion

We have now all the elements to construct a global analysis. A general model is nevertheless difficult, for substantial as well as technical reasons.

The first reason is the one of structure. Two different approaches were mentioned: in one sense, we can suppose that, controlling by social position and urban representations, use of public spaces is a function of socio-political insertion. In other words, it is because an individual has subjectively found his place in the society that contacts with other will be frequent. However, the alternate model could also be proposed: that contacts with others, in the public spaces, will determine relation with the urban society. We have there the already mentioned difference between defined space and determinant space.

Figure 7.5 Use of the three public places as a function of social
position, age, gender, and nationality

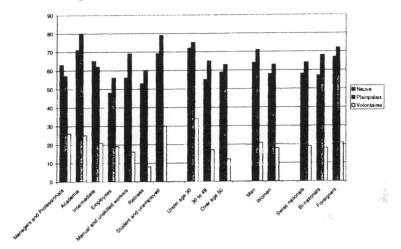

From a technical point of view, we have developed some mul-
tivariate models based on multiple regression and generalised linear models which are too complicated to present here in detail, but
we will try to present the essence of these results. Even if the statistical explanation is not always as good as a researcher might want,
some interesting points appear. It should be mentioned that all the
variables used here have some impact in one or another part of the
model, but not all the indicators appear in all the equations. This
selection of variables is also a point of interest: to show that use of
public spaces is independent of the perception of conflicts in the
city is as important as showing that this is a function of social position.

Defined space. What, then, are the variables which best explain the
utilisation of public places? Or, more precisely, what are the variables which take into consideration the use of the three public
places studied and of the city centre? In other words, it is the route
of the «defined space» which we are following here.

For the three public places mentioned, the variables describing
social position play an important role: first of all, use declines with
age. It should be noted however, that two indicators reveal opposing results: the higher the income, the less use is made of the place.
Simultaneously, and if taken equally, the higher the education, the
more use is made of the place. This result is important for two reasons: first, it shows that the same social positions do not necessarily entail the same degree of use of a public place. Although a

connection certainly exists, it does not place all individuals on an equal footing. The less intensive use of public places made by the most wealthy residents is telling in this context, *since* it is well known that urban segregation is first and foremost a characteristic of the privileged classes, along with the corresponding issue of solidarity: today in Switzerland, the indices of segregation are higher for the privileged social categories than for the less privileged.

Figure 7.6   Proportions of respondents who say they appreciate the three public places according to use and perception of diversity

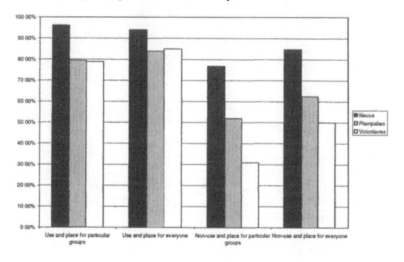

The years spent in education have an opposite effect, which evokes the debate over the reappropriation of the city by those social categories having the greatest cultural capital. The literature of the last decade on new social movements in the urban environment has highlighted this differentiation between economic and cultural capital. This is undoubtedly the affirmation of an urban lifestyle with particular characteristics. The credibility of this supposition increases if one notes its link with the fact that numerous activities are carried out in public places. It should also be noted that if the image of the city is excluded from the use made of it, the relationship to others, which is the second aspect of social integration, has an influence. Furthermore, trust in authorities also has an impact, through an opposite one: the greater the overall trust, the less intensively public places are used. This finding reflects a portrait of a category of the population, the relatively young, critical and well educated group who have a positive image of others in horizontal relationships, but do not manifest trust towards authorities.

The portrait changes completely when the use of the city in general is described, that is, without considering the use of the three places which we just mentioned. In this case, if social position has an influence here, it is only partial and does not involve any impact of age and education. This result confirms the observation of a city centre that is used by the entire population. This somewhat utilitarian use of the city centre can be deduced because the variables defining participation and integration are almost devoid of influence. On the other hand, symbolically speaking, the affirmation of urban qualities also plays a small though nevertheless statistically significant role. Finally, although activities in public places serve to explain the use of the city centre, private activities also have an influence. In this context, use of the city centre is rather linked to daily, or even functional, activities, instead of an image of the city based on social value judgements.

Determinant space. The problem here is more complex: we can interest ourselves as to the influence of use of public space on social insertion as well as on political participation. Moreover, each of these concepts has multiple aspects, ones linked to local questions and other to the society space.

a) Social integration: Social integration as we have measured it comprises two aspects: integration within the neighbourhood, and the relationship of trust with respect to the whole city population. In its first aspect, which focuses on the local, social integration can hardly be explained by the variables selected: the model is capable of explaining only 5% of the variance. It is nevertheless interesting to comment on those variables which have an influence, as well as on those which have none. The local aspect is centred around gender: women would appear to be integrated at a more local level, but no other variable describing social position plays a significant role. In other words, what we may interpret as a position of withdrawal into an intermediary sphere encompasses all social layers. Furthermore, this type of integration is also characterised by a perception of society as potentially conflictual, but this perception is a function of demographic groups rather than of social classes. This vision of group conflict nevertheless does not rule out an attitude of trust in authorities nor a traditional form of political participation, compatible with local social integration. The affirmation of a very localised integration does not result in wider integration into the networks functioning at the city level, nor does it generate greater tolerance towards other groups; this type of integration is accompanied by a positive attitude with respect to the systems governing the city. For integration into the urban environment as a whole, the statistical explanation is much better. All things being equal, integration increases with one's social position, whether measured in terms of income or education. Similarly, activities outside the home increase the possibility of a positive relationship to the city. Other variables also have an influence: for example, when social activities are considered to be of an urban quality, the chances are greatly increased that the respondent will be well integrated in this domain. In the same vein, the more that human elements are seen as disturbing factors in the city centre, the less likely this form of integration is to exist. Finally, all three types of political participation – local, traditional, and delegation to authorities – tend to increase urban integration. A final element in this area: use of public places tends to reinforce this integration to a certain degree. However we find a set of variables which would appear to support the hypothesis of an urbanness comprising the use of public places in addition to political participation and social integration.

b) Socio-political participation. What we have referred to as resident participation is undoubtedly the variable that is the least clearly explained by the model. This is not surprising, since the degree of direct participation of residents does not depend on their social category. Although it is also not surprising to have no link between this participation and the use of public places, whether these are comprised in the three places selected or in the city centre, it should be emphasised that this participation is independent of the image which one may have of the city. In other words, this type of local participation tends to be self-contained, and to disregard any links that may exist with other levels of participation outside the neighbourhood. The only exception is that social integration which is urban has somewhat of a tendency to increase the significance attributed to this form of local participation. Contrary to inhabitant participation, conventional political participation is better explained. Once again, this comes as no surprise when one considers that this type of participation is deeply ingrained in the different social layers (Roig, 1975, Kriesi, 1993a). Nor is it surprising that this type of participation is not related to the perception of the city. On the other hand, close local and urban integration tends to reinforce conventional political participation. This result supports the idea that political participation is closely linked to social integration.

The third form of participation which we have retained is somewhat peculiar, since it entails delegating to authorities. Three findings are of particular interest in this context. First, this type of delegation increases with age and income, but decreases with education. This therefore reflects an opposing factor to the social movements present in urban areas, which are rather characterised by a cultural capital greater than the financial capital (Kriesi, 1993b). Second, trust and delegation are more likely to occur if the image of society is relatively free of conflict, which nevertheless does not translate into a shared vision of urban life: in this case, social groups are seen as the elements most likely to disrupt urban harmony. Third, with respect to politics, this attitude is linked to social integration, whether local or city-wide. These results coincide with the idea that the best integrated groups may experience a relationship of trust with the system without necessarily manifesting greater tolerance.

Conclusion

In conclusion, we should first note that public places do indeed provide clear indicators to be used in the analysis of the urban environment today. Their role is important for residents as well as for authorities; an analysis of the use of these places reveals that they are part of a problem concerning all issues of urban life: in particular, pure appreciation of urban characteristics, socio-political integration, and relationships with others. What is the relationship between public places, urbanness, and integration? Does the use of public places mean that participation and exchange will necessarily follow? More exactly, can a choice be made between a hypothesis which portrays integration as part of a complete picture that includes the use of public places, and the reverse hypothesis which holds that the use of a public place is a factor of integration?

One comment to be made is that there are two components to be highlighted when referring to the use of public places: in the city centre, there is a certain homogeneity of characteristics and social value judgements; Geneva's centre is seen as a space open to all, a space which everyone feels entitled to use, see as part of the common heritage and is viewed as part of the city's wealth of attractions. It can be noted at the very most that anyone with leisure activities, whether anonymous or involving a group, who appreciates the qualities of the city, has greater chances of making use of the city centre. However, since this use made of the city centre is primarily functional, one may wonder whether it has the potential to contribute to increasing urbanness. On the other hand, the specific public places located on the fringes of this centre have a more diverse social profile; the actors of social movements who are young and possess a high measure of cultural capital are over-represented. Social integration also tends to have a positive effect on the use of the public places under consideration. In this context, one can effectively speak of "defined spaces".

The hypothesis of "determinant spaces" cannot be entirely ex-cluded either. Although it is true that the use of these places is still a variable associated with social position in the analysis, the per-ceptions of the city and the groups comprised within it are also relatively independent of the use of specific public places neverthe-less, relationships of trust with others or with authorities are gen-erally stronger in the proportion to the increased use of public places. In this context, it can be said to a certain extent, that the public places studied create urbaneness. A certain number of res-ervations must be made with respect to this portrait. First, social integration as well as political participation cover multiple facets. In particular, there is a very local dimension which is related nei-ther to urban issues nor to community living. Hence for public places to be used, it is not sufficient to be integrated into society, or even into one's neighbourhood, nor to appreciate the diversity of the groups and expressions that make up the city. On the contrary, a positive attitude towards the action of local authorities, all other things being equal, is directly related to the use of public places. At the same time however, this does not necessarily imply acceptance of the others: there is greater probability that other groups of users will be cited as the negative elements in the city rather than more objective nuisances such as noise or pollution.

This also means that although public places are important to the functioning of a city, although they are important in the appre-ciation of the compact city, developing such places artificially or artistically is not sufficient to ensure that such places will "work" at the wave of a magic wand.

Public spaces and the new urbanism

In Europe, the debate on public spaces has been intense in recent years, precisely with the idea to increase life quality and city attraction. In the United States, there was also a reflection on the problems caused by urban sprawl and pollution associated with private transportation systems. An important movement in this context is the new urbanism or the research of an urban form, where, like the city of past centuries, public spaces and meeting places are important, where urbanization will be dense and constructed around real streets. This is supposed to have some consequences on living conditions. As Lehrer and Milgrom point out: "One of the new urbanism's main assertions is that the resurrection of an abandoned urban form will help recreate a lost idealized type of community, in other words, that community follows form" (1996, p. 50).

Such a proposition raises questions that we have indirectly addressed in the companion chapter:

How, and to what extent, is it possible for urban form to induce social behavior, particularly in such key issue as the creation of a community? Data indicate that there is no automatic relation and, if any, that the direction of influence is from the social behavior to environment rather than the reverse.

What about the idea of community? Is it a really egalitarian one? It is true that community reinforces the links between members of the concerned group but, in this case, it tends to exclude even more strongly the people outside the community. Such a movement can be even stronger as the housing type produced by the new urbanism is more likely to be dedicated to the middle classes rather the poorest ones.

Another point, in relation to the question of the environment's control by the people living in a neighborhood or, more generally, an urban unit, is the real status of the public spaces. In many cases, publics spaces are no more really public but, in fact, private or semi-private places. A classic example of this, is the area around or inside commercial centers, where a private organization partially carries out the police's function and can choose who is allowed to use these semi-public spaces. This is important for the daily life of the inhabitants, also in European cities, were more and more people use these spaces as places where they can simply loiter, even if the weather is not so favorable.

Dominique Joye

References

ADRIAANSENS, Ans. (1994) "Citizenship, work and welfare." Pp. 66-75 in *The Condition of Citizenship*, edited by Bart VAN STEENBERGEN. London, Thousand Oaks and New Dehli: Sage.

BELLAH, Robert N. (1997) "The Necessity of Opportunity and Community in a Good Society." *International Sociology* 12:387-393.

BIAREZ, Sylvie. (1998) "Territoires, espaces urbains, espaces publics. Une approche de l'action publique locale en France." *Revue Suisse de science politique* 4:67-89.

BLANC, Maurice (Ed.) (1992) *Pour une sociologie de la transaction sociale*. Paris: L'Harmattan.

BREHM, John, and Wendy RAHN (1997) "Individual-Level Evidence for the Causes and Consequences of Social Capital." *American Journal of Political Science* 41:999-1023.

CATTAN, Nadine, Denise PUMAIN, Céline ROZENBLAT, and Thérèse SAINT-JULIEN. (1994) *Le système des villes européennes*. Paris: Anthropos.

ETZIONI, Amitai (1995) "Old Chestnuts and New Spurs." Pp. 16-34 in *New Communautarian Thinking*, edited by Amitai ETZIONI. Charlottesville and London: University Press of Virginia.

GÄCHTER, Ernst (1998) *Stadt Bern: Einwohnerbefragung 1997*. Bern: Statistikdienste der Stadt Bern.

GAUDIN, Jean-Pierre, Philippe GENESTIER, and Françoise RIOU (1995) *La ségrégation: aux sources d'une catégorie de raisonnement*. Paris: Plan construction et architecture.

HAMEL, Pierre (1998) "Urban Policies in the 1990s: The Difficult Renewal of Local Democracy." *International Political Science Review* 19:173-186.

HAUMONT, Nicole, and Jean-Pierre LEVY (Eds.) (1998) *La ville éclatée, quartiers et peuplement*. Paris: L'Harmattan.

HIRST, Paul (1998) "Vers la démocratie associationniste." *Recherches, la revue du Mauss semestrielle* 11:168-174.

INGLEHART, Ronald (1990) *Culture Shift in Advanced Industrial Society*. Princeton: Princeton University Press.

JOYE, Dominique, Thérèse HUISSOUD, and Martin SCHULER (Eds.) (1995) *Habitants des quartiers, citoyens de la ville?* Zurich: Seismo.

JOYE, Dominique, Jean-Philippe LERESCHE, Martin SCHULER, and Michel BASSAND (1990) *La question locale, un éternel sujet d'avant-garde?* Berne: Conseil suisse de la science.

JOYE, Dominique, and Martin SCHULER (1995) *Stratification sociale en Suisse: catégories socio-professionnelle*. Berne: Office fédéral de la statistique.

KAMINSKI, Antoni Z (1991) "The Public and the Private: Introduction." *International political science review* 12:263-265.

KRIESI, Hanspeter (Ed.) (1993a) *Citoyenneté et démocratie directe. Compétence, participation et décision des citoyens et citoyennes suisses*. Zurich: Seismo.

KRIESI, Hanspeter (1993b) *Political Mobilization and Social Change*. Avebury: Adershot.

KÜBLER, Daniel, Dominique MALATESTA, and Dominique JOYE (1997) "Une politique locale à l'épreuve des faits: conflits de localisation et services urbains." Pp. 254-269 in *Gouvernance métropolitaine et transfrontalière*, édité par G. SAEZ, J.-Ph. LERESCHE, and M. BASSAND. Paris: L'Harmattan.

LAMARCHE, Hugues (1986) "Localisation, délocalisation, relocalisation du milieu rural." Pp. 69-99 in *L'esprit des lieux*, edited by Programme d'observation du changement social. Paris: Editions du CNRS.

LECA, Jean (1997) "Préface." in *Le gouvernement des villes. Territoire et pouvoir*, édité par Francis Godard. Paris: Descartes & Cie.

LEVI, Margaret (1996) "Social and Unsocial Capital: A Review Essay of Robert Putnam's Making Democracy Work." *Politics & Society* 24:45-55.

LEVY, Rene, Dominique JOYE, Olivier GUYE, and Vincent KAUFMANN (1997) *Tous égaux? De la stratification aux représentations*. Zurich: Seismo.

LOFLAND, Lyn (1993) "Urbanity, tolerance and public space. The creation of cosmopolitan." in *Understanding Amsterdam*, edited by Léon Deben, Willem Heinemeijer, and Dick van der Vaart. Amsterdam: Het Spinhuis.

LOFLAND, Lyn H (1998) *The Public Realm*. New York: Aldine De Gruyter.

MAY, Nicole, Thérèse SPECTOR, and Pierre VELTZ (1998) "Introduction." in *La ville éclatée*, edited by Nicole MAY, Josée LANDRIEU, Pierre VELTZ, and Thérèse SPECTOR. La Tour d'Aigues: L'Aube.

MAY, Nicole, Pierre VELTZ, Josée LANDRIEU, and Thérèse SPECTOR (Eds.) (1998) *La ville éclatée*. La Tour d'Aigues: L'aube.

MCKAY, David (1996) "Urban Development and Civic Community: A Comparative Analysis." *British Journal of Political Science* 26:1-24.

O'CONNOR, Justin (1997) "Donner de l'espace public à la nuit." *Annales de la recherche urbaine* :40-46.

OFFNER, Jean-Marc, and Denise PUMAIN (Eds.) (1996) *Réseaux et territoires, significations croisées*. La Tour d'Aigues: L'aube.

PLAN, Commissariat Général du, and the Rapport du groupe lead by Jean-Paul Delevoye (1997) *Cohésion sociale et territoire*. Paris: La documentation française.

POLLIT, Katha (1998) "For Whom the Balls Rolls." *The Long Term View* 4:14-15.

REMY, Jean, and Lyliane VOYÉ (1981) *Ville, ordre et violence*. Paris: Puf.

ROCHÉ, Sébastien (1998) *Sociologie politique de l'insécurité*. Paris: Presses universitaires de France.

ROIG, Charles (1975) "La stratification politique." in *Les Suisses et la politique*, edited by Dusan Sidjanski. Berne: Lang.

SCHNAPPER, Dominique (1998) *La relation à l'autre. Au cœur de la pensée sociologique*. Paris: Gallimard.

SENNETT, Richard (1982) *La ville à vue d'œil*. Paris: Plon.

SMITH, D.M., and Maurice BLANC (1997) "Grass-roots democracy and participation: a new analitical and pratical approach." *Environment and Planning D: Society and Space* 15:281-303.

TIÉVANT, Sophie (1983) "Les études de communauté et la ville : héritage et problèmes." *Sociologie du travail* 83.

8 Participation and Quality of Life: Experiences with Local Agenda 21

LUDGER BASTEN AND LIENHARD LÖTSCHER

Introduction

The current debates on life in the city leave a confusing picture.[1] There are many discourses, a plethora of themes and a plurality of theoretical starting points. Where applied urban research is concerned, however, theoretical complexity tends to take a back seat as issues of planning and urban politics focus on overriding themes. On a day-to-day level, the problems and issues confronting cities in the developed world seem to be rather universal: unemployment, social disintegration, the need to upgrade and modernize infrastructure for economic development, increasing transportation requirements, worsening problems of waste and environmental management. It seems, though, that two themes have gained increasing prominence over the last five to ten years: the first is the question of quality of life, the second is generally described by the term "sustainability".

While the former theme is not altogether new, the debate on urban quality of life today is less concerned with a critique of modern, supposedly rational urban planning projects and their uneven distribution of costs and benefits. Rather, the concept now assumes many heterogeneous urban societies with differing views of quality of life thus making process issues of involvement and fair participation more prominent than material issues. In other words, communication about the use of common urban spaces, urban politics and the whole realm of civic life have become aspects of the

---

[1]   We would like to thank Markus Raser for extensive research assistance for this paper.

concept of quality of life itself.

The "sustainability" concept has developed from older environmental debates on an international level. However, it stresses the realization that environmental sustainability is invariably intertwined with the spheres of economic and social sustainability.[2] In this line, the 1992 United Nations Conference in Rio de Janeiro developed an agenda for the 21st century ("Agenda 21") containing two important aspects relevant to urban development. First, it established sustainable settlement (and thus urban) development as a distinct goal. Second, it focussed specifically on the local political sphere as a central arena for implementation by calling for the creation of a so-called "Local Agenda 21" (LA 21) in municipalities around the world. This was to be established through a broad consultation process with a multitude of local actors and civic groups rather than to be developed just by the local political-administrative systems (Bundesministerium für Umwelt, Naturschutz und Reaktorsicherheit, n.d.).

These two concepts, "sustainability" and "quality of life" are both inherently procedural, and, moreover, both tend to unfold primarily at the local level. We would argue that this is central to the understanding of current urban change in Europe.

Notable changes have occured in the urban political sphere. This can be illustrated by the discourses about the concept of the local state (Duncan and Goodwin, 1988) which connects local political developments with global processes of structural economic change. The presumption is that national governments are increasingly losing the power to influence or direct economic and societal development - largely due to processes of economic globalization (Wallace, 1992) which produces a heightened and global inter-urban competition for investments and jobs.[3] This leaves the local (political and economic) arena as the scene where positive and negative effects of economic restructuring and socio-cultural developments become most visible (Dangschat, 1996). It is here that conflicts between interest groups are being fought out as local resources get scarcer and while the national political system tends to withdraw central control. Hotly disputed, however, is the clout that the local state has in dealing with the range of problems surfacing in the cities. While some writers see it as even less powerless

---

[2]   For the emergence of sustainability as an issue for international debates on development and the environment see the "Brundtland Report" (World Commission on Environment and Development, 1987).

[3]   The conventional wisdom now stresses cultural and natural amenities as prime factors of location - their connection to quality of life and environmental/sustainability issues should be fairly obvious.

than the nation state vis-à-vis the flexibility and influence of global capital, others discover new opportunities for democratic, decentralized politics as urban areas are left to themselves to work out how they want to do business and how they want to envisage and shape their own futures (Krätke and Schmoll, 1987; Mayer, 1991)Local political agendas, then, are not just small-scale versions of national ones. Rather, the imminence of problems suggest a sharper focus on issues of quality of life. Disputes and conflicts most often surface when effects of specific projects are felt or envisaged by the people and communities directly concerned. Not surprisingly, mobilization for citizen participation and direct action work best on a local scale. In fact, the conflicts that partly stem from diverging definitions of quality of life demand a more participative approach to urban politics, and only the local level seems to be able to provide an arena for this kind of participation.

Urban development, planning and participation

Quality of life and sustainability are certainly not new themes in urban planning, even if their importance may have increased. However, analyzing their articulation and negotiation, one finds that new forms of dealing with these themes have emerged.

Planning systems of the past

Popular participation in planning is largely a child of the late 1960s and 1970s. Before that, planning had largely been seen as a profession: the art, science and craft of plan-making to be carried out by professional planners (Albers, 1988). In this sense, it had been separate from political decision-making. Planners were bureaucrats employed by local or regional governments, while politicians and the politics of planning decisions had been defined as being outside the realm of planning itself (Lötscher, 1985).

In the late 1960s, perceptions were shifting as an array of protest movements caused substantial debates on political participation in democratic societies, leading to a more informed and politically active public. Planning was now starting to be viewed as a larger process, with political decision-making - and thus elected politicians - being part and parcel of the whole process. Another major shift in perception was the acceptance of the public at large as a separate group of actors (Lash, 1977). It involved the idea that it was not only planners who would be deciding what was best for the common good. Participation soon became one of the central points of prescriptive planning theory (Dienel, 1978).

Two important aspects of this challenge to the existing planning system should be stressed. First, the citizen groups that led the challenge were almost invariably protest groups rallying *against* proposals for urban development or modernization.[4] Politicians and planners were their opponents, since they were seen as the active agents of urban change. Second, private business, especially the development industry (Lorimer, 1978), with its intrinsic interest in construction, (re-)development and renewal, was starting to be recognized as a separate and powerful group of actors. The heterogenity of the public at large became notable.

The response of the political-administrative system was integration: participation was established as part of the planning process, planning laws were changed, planners adopted by holding public meetings and trying to mediate between often conflicting interests. And yet, the system itself remained intact, since there existed a common belief in the potential of science and comprehensive, systematic management. Modifications to the planning system mirrored both the euphoria and the rational, comprehensive approach that would benefit from more citizen input.

New planning systems?

Over the last 15 years, the two optimistic cornerstones of urban planning, the euphoric belief in a quasi-natural law of (material) progress and the belief that social systems could be rationally analysed and therefore planned for, have largely been eroded. The failure of many redevelopment projects, the persistence of grave social problems and a general trend towards deregulation have again led to significant modifications of urban planning systems (Hall, 1988). They concern the constellation of actors, the contents or issues and - our central argument in this paper - the whole process-context of planning.

Old and new actors. The politics of urban development today seems to be a more open process than in previous decades. One important cause of this is doubtlessly the increasing scarcity of public funds. Meagre economic performance has eroded many municipal tax bases, while the fiscal requirements of social security programmes

---

4   These groups would try to protect their properties, neighbourhoods and lifestyles, often discovering and/or using more general protest issues like the environment, architectural or historic preservation. What emerged was an interesting mix of conservative motives and progressive politics.

have tended to increase. Furthermore, costly failures in the past have reduced many planners' confidence in the possibilities of planning as such, while leading to the withdrawal of national pro- grammes and funding. Without being able to control or access pub- lic (i.e. government) funds for development, politicians and plan- ners could produce plans but the investments needed to realize such projects were often lacking. Not surprisingly, then, these two "classic" groups of actors have had to devise a new approach to devising and practicing urban development and planning.

One common solution has been the search for private funds. With less government funds available, municipal actors have more actively sought out private investors rather than just waiting to act on proposals (Basten, 1998). This has effectively produced a wider, more diverse circle of actors, since private business interests, inves- tors and developers are now being seen as partners in the whole process rather than as petitioners (Drescher and Dellwig, 1996; Selle, 1994).[5] Hence, they become active participants in the negoti- ating process early on, thereby increasing the number of actors involved.

However, the scarcity of public funds and the growing impor- tance of new factors of location have also led to the involvement of new government actors. As attracting outside investment for urban development has become a strong priority, other areas of munici- pal government have been forced to re-orient their policies accord- ingly (e.g., in the provision of cultural infrastructure, parks, recrea- tion and leisure facilities) (Harvey, 1989). Accordingly, urban de- velopment now involves not only planning in a narrow sense of the word (plan-making and engineering) but also other municipal activities that can influence city image and marketing (Helbrecht, 1994). Furthermore, the global character of inter-urban competition has increased the necessity of a regional approach, often bringing different tiers of government into the game (Selle, 1994).

It should not be overlooked that the general public has also become more actively involved in issues of urban development. On the one hand, this is another result of municipal governments' search for partners. When it comes to projects supporting or im- proving quality of life - parks, flower shows, (cultural) festivals etc. - particular citizen groups can often be called upon to work with municipal administrations hoping for positive, image-promoting effects. On the other hand, many such projects also encounter resis-

---

5   The term "public-private-partnership" illustrates this changing approach and
       perspective (Heinz, 1993). Under such arrangements, governments and pri-
       vate actors get together to pool financial and manpower resources or know-
       how for mutual benefit - though the evaluation of public benefits is often
       controversial.

tance from other citizen groups, reminding us of an increasing differentiation of urban societies and a pluralisation of interest groups. In actual fact, many such groups have formed as a result of the retreat of the state. These are often intensely critical of or in opposition to municipal (or even national) government policies, trying to establish alternative or self-help projects in fields such as employment policy, youth policy or environmental affairs). This differentiation of the general public has produced a range of new potential actors, and their integration into the process has obviously become a far more difficult and complex task.

New processes. With a greater variety of actors involved, and with the "classic" actors (politicians and planners) having lost much of their former power, it is only logical that the processes of urban planning have changed considerably as well. In fact, for Selle (1994) the procedural aspects of the new culture of planning are fundamental. He chooses the term cooperation to emphasize its central characteristics, while Healey (1992) writes of "the communicative turn in planning theory". For these authors, planning has become an exercise in facilitating communication and cooperation among many diverse actors and working towards a consensus on development issues - be they specific projects or plans and programmes. This means a significant role change, since it emphasizes negotiation and communication rather than setting agendas. This requires new skills and new forms of organization very different form hierarchical bureaucracies: networks, round tables, conferences, workshops etc. Though hardly new, such forms of organization take on a more dominant position in the way planning is being done. At the same time, processes tend to be more open than in traditional planning, since the old actors have lost much of their dominance for setting and pushing through a particular agenda. In general, a more project-based approach also follows, since the motivation to get involved increases significantly if issues are specific and hands-on rather than abstract and long-term. As rough as this sketch of changing processes may be, the need for actors to adapt to them is obvious.[6]

Many administrative systems of urban municipalities have proven too rigid to deal with such open, fluid communicative environments. They require flexibility, openness, a decentralisation of decision-making powers and speed. Similarly, such processes require all relevant experts to be involved early on, which favours

---

6   There is no doubt that much of planning still follows a more regulative approach and that more traditional techniques still abound (Selle, 1994, p. 43 ff.).

the setting-up of interdepartmental project teams rather than traditionally segmented departmental structures (Drescher and Dellwig, 1996). Not surprisingly, then, government reform at the municipal level has mirrored these developments in many municipalities across Europe (Banner and Reichard, 1993).

But private business interests and developers have also had to adapt to the changing scene of urban development. On the one hand, through early negotiations, they can reduce planning risks and thus save costs and time. On the other hand, system flexibility has also increased uncertainty regarding the outcome of such negotiations. Projects may need to be modified, as local authorities push their own goals, requiring private businesses to be more open and flexible. More often than not, though, case-based and sometimes tailor-made planning decisions can produce substantial advantages for the private sector.[7]

However, the flexibility of the negotiating process has also made it potentially more open to the influence of other interest groups. While many developers initially seek confidential negotiations with the authorities, the relative openness of the process invariably leads to the involvement of such interest groups, often supported by municipal authorities committed to public participation or just proactively avoiding potential protests. In this way, private non-business interests have become involved as actors in processes from which they had formerly been excluded. And with better technical means of getting organized and influencing public opinion, they gain a stronger voice in many planning decisions.[8] Naturally, developers and business-interests have adapted in kind, with marketing and public relations campaigns becoming an integral part of any major project proposal that touches on important issues for the communities involved. While this has not necessarily led to better or more rational planning decisions - or to a notable shift in the relative power of the actors involved - there can be little doubt that planning for urban development has generally become a more public process in due course.

A new stage. The variety and heterogeneity of these new actors and their increasing involvement in urban development has attracted

---

[7]   It is not surprising that such practices (e.g., discretionary zoning) have sometimes caused severe protests by the general public or some special interest groups. These warn against the dangers of corruption and of a "sell-out" of public interests as planners and developers cooperate throughout the planning process.

[8]   Consider the relative ease of producing issue-based newsletters or protest flyers through the widespread availability of computers and DTP-software, or of researching information through the internet and establishing websites or discussion forums.

the attention of researchers over the last few years. Since many such actors have formed around issues and problems once seen as largely the task of governments, they are often neither private nor public, but a mixture of both. They have variously been labeled "third sector", "intermediate sector" or "intermediate organizations".[9] Our focus here, however, is not primarily on the actors themselves, nor on the way they have organized to get involved. Rather, it is the "stage" on which the actors play out the drama of participation in urban development today.

Selle (1994) sees this "intermediate sphere" as the place of cooperation, Friedman (1987) talks more loosely of the "public domain". But whatever to call it, we would argue that this stage, on which negotiations about the present and the future city take place, is the central element of the new politics of urban planning and development. It is a stage that doesn't belong to any one or any group of actors. Rather, it is where all the different actors meet.[10] Participation now implies that the stage is wide open for new initiatives, which may be started by virtually anyone, and that the agenda for urban development can no longer be predetermined by the traditional powers. These procedural aspects of participation are closely linked to quality of life.

Local Agenda 21 in Munich: an illustrative example

As we discussed above, the issues of quality of life and sustainability have gained increasing prominence in the discourses on urban development in recent years. In some way, these priorities already reflect the new stage of urban development planning, since the international debates on the state of the environment have been very actively and successfully influenced by environmentalist lobby groups that can serve as a prime example of those "intermediate organizations". These organizations, together with equally successful groups in the field of international development, also heavily influenced the course and the agendas of the 1992 Rio conference which culminated in the Agenda 21 and its chapter on local action. Hence we have chosen to investigate this new stage by analysing the current debates and processes of urban development as contained within the framework of a Local Agenda 21. The city of Munich was one of the first municipalities in Germany where an

---

9   Selle (1994) rightly points out that these "intermediate organizations" are not
        all new, and that several countries have long-standing traditions of some
        such groups and organizations.
10   To make cooperation possible, sometimes new actors have appeared specializing on facilitating negotiations and, if need be, arbitration.

LA 21 process was started and thus allows for a case study of some depth.[11]

A chronology of LA 21 planning in Munich

The history of planning for a Local Agenda 21 in Munich started in the spring of 1994, when programme planners of the Münchner Volkshochschule (MVHS), the local adult education institution, chanced upon the results of the Rio conference (Kreuzer, 1998a).[12] What appealed to them was the specific call for the development of local agendas in chapter 28 of the final Rio document (Bundesministerium für Umwelt, Naturschutz und Reaktorsicherheit, n.d.).

Since "sustainability" and "Agenda 21" were terms not yet widely known among the general public, they decided to start a discussion process with local initiatives and organizations. In April 1994 representatives from various Third World initiatives, environmental organizations, citizen participation groups, trade unions, churches and political parties were invited to debate the starting of an LA 21 process for Munich. Out of this emerged a loose forum of some 30 local initiatives and organizations which spent nearly a year drawing up a multi-layered strategy to develop a participatory LA 21 process. By the autumn, MVHS was starting a programme of courses, workshops and lectures to spread information about the ideas of Rio 21 and about the project to develop an LA 21 for Munich.

---

[11]   For this case study, we conducted a number of telephone interviews with various participants in the LA 21 process in Munich (a list can be obtained from the authors) as well as extensive document research. We have refrained from referencing individuals, partly because information was repeated several times, partly to maintain anonymity as far as possible.

[12]   There is a system of "Volkshochschulen" (literally: people's high schools) all over Germany. They are government-funded and tend to run a very diverse programme of courses, lectures, seminars etc. that are open to everyone for a small fee.

Figure 8.1 Key NGOs involved in local agenda processes in
Munich. Source: material supplied by Munich agenda
office (modified by authors)

ROBINWOOD
Action Group for
Environment & Nature

GLOBAL DENKEN
VERNETZT HANDELN
Global Challenges Network

GREEN CITY

Ecomobility Network

Ludwig Bölkow
Systemtechnik
Eco-technical
System Research

PLANUNGSGRUPPE
City Planners

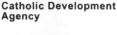
Research Organization
on Social Innovation

MISEREOR

Catholic Development
Agency

NORD
SÜD
FORUM
Münchner e.V.
Intercultural Relations

MÜNCHNER AGENDA 21

Initiative
"Mehr Demokratie in Bayern"
Direct Democracy
Initiative

Die
Umwelt-
Akademie
Enviroment Academy

Bund Naturschutz
in Bayern e. V.
Kreisgruppe München
Alliance for the Protection
of Nature

IMU-Institut
Institute for Media- and
Urban Research

MÜNCHNER
FORUM Münchner
Diskussionsforum für
Entwicklungsfragen e.V.
Intermediate Discussion
Forum for City-Development

Münchner
Volkshochschule
Institute for Adult Education

CHILDREN FOR A BETTER WORLD
International Network
for Children Rights

Evangelisches
Forum
München
Adult Education of
the Protestant Church

Pro REGENWALD
Initiative for Tropical
Rainforest

In March 1995 an official opening event took place, jointly organized by MVHS and the forum of initiatives. By then, the local
political system had also become involved: the lord mayor became
patron to the opening event, and in May 1995 city council unani-
mously decided to have an LA 21 developed for the city. Council

delegated this responsibility to the department for environmental protection, charging it with the specific task of setting up an LA 21 process under the umbrella of a broad and participatory "forum sustainable Munich". In this way, city administration, MVHS and local initiatives joined forces to devise a structure of organization and a plan for future activities. One and a half years were thus spent establishing communication between the different actors involved and to develop a concensus about goals and the steps required to reach them. The early focus was on finding a structure of organization and on establishing mechanisms for consultation and cooperation.

By spring 1996, the basic structure of organization had been agreed upon, with a co-ordinating office and four theme-based special committees. These organized a multitude of workshops, conferences and round tables to involve the public and develop specific projects. Over the summer of 1997, an extensive activity campaign was organized to increase citizen participation and to decentralize action down to the level of individual neighbourhoods (David and Bock, 1998). This extensive consultation phase was eventually to lead to a comprehensive Local Agenda 21 document by the spring of 1998. Since then, the LA 21 process has undergone significant changes in its organizational structure and focus of work.

Organizations and structures

During the consultation phase, from 1996 to spring 1998, the organizational structure of the LA 21 process in Munich consisted of five main entities: the agenda office, the steering committee, the advisory board, the citizens' forum and the four special committees.

The agenda office was the central process management unit. Staffed and financed as a kind of public-private-partnership by the department for environmental protection, the planning department and the (private) Global Challenges Network, the office functioned as the hub of co-ordination within the LA 21 process. It organized budgets, disseminated information to groups and civic departments, prepared meetings and conferences, documented and evaluated the process at large and took charge of media campaigns and public relations. There were some seven to ten staff in the agenda office, three of whom were seconded from civic administration.

**Figure 8.2   Structure of organizations of LA 21 planning in
                 Munich - consultation phase**

Source: Eckardt, 1997, p. 22

The steering committee oversaw the LA 21 process as a whole. Its central task was to forge a comprehensive LA 21 programme out of the work of the special committees, the advisory board and the citizens' forum. It thus consisted of some ten representatives of the other four LA 21 entities and met four or five times a year.

The advisory board consisted of some 40 local personalities supporting the work of the special committees through expert advice and guidance. They had been invited to the board by the steering committee acting on nominations from the agenda office, local initiatives or individuals. The advisory board met nine times and conducted two workshops debating project proposals with the respective special committees. The results of these consultations were published in a report to council and to the public at large.

The citizens' forum was open to everyone wishing to join individual meetings or lectures. It was set up in the spring of 1995 through MVHS and emerged out of the loose forum of initiatives and organizations that had initially started the LA 21 process. Its aim was to involve as many individuals and groups in the debates about LA 21 planning as possible through debate and to suggest topics or even projects to the special committees. Monthly meetings with invited experts usually drew about 40 people per meeting. Its activities were planned by a working committee consisting of representatives of four separate initiatives including MVHS. Not surprisingly, then, the citizens' forum activities were closely tied to the adult education programmes of MVHS, who created a separate series of lectures and courses entitled "Sustainable Alternatives for Society, the Economy, Science and Politics".

Lastly, there were the four special committees which carried out most of the actual work on the main issues of sustainable urban development. It was here that the general concepts and issues were broken down and that projects were drawn up. Each committee comprised some 30 individuals who met about eight to ten times a year. These special committees showed great flexibility in terms of internal organization and methods, using working groups, round tables, scenario workshops and a variety of highly communication-based planning techniques. In each, a five-person co-ordinating team was in charge of organization and co-ordination with other LA 21 entities and municipal departments. Committee members were selected by the co-ordinating team, based on candidates' know-how, committment and creativity after a public call for nominations. The following four special committees were formed:

Figure 8.3 Project work and communication-based planning methods: a workshop scene

Source: photograph by Oliver Mayer

The Special Committee on Employment and Economic Development sought to develop strategies and projects for sustainable economic and labour market structures. Its projects focus on issues of sustainable production techniques and environmental protection as well as on new forms of employment and community work.

The "One World" Special Committee set itself the goal of bringing the developing world closer to Munich. It dealt with a range of issues from fair trade and international investment to the integration of ethnic minorities into the local community and spreading one-world education in schools.

The Sustainable Life-styles Special Committee concentrated on the effects of daily activities and consumption patterns. Trying to establish life-styles that can save resources and prevent pollution, the committee focussed on the level of individuals' daily behaviour to find pathways to a healthy, ecologically and socially sound environment.

The Special Committee on Housing, Settlement and Mobility examined the most problematic issues of urban planning in a narrower sense of the word: transport, housing, infrastructure and (sub-)urban expansion. It was arguably the committee with the most diverse membership, spanning the full range of the political spectrum, chambers of industry and commerce, environmental protection groups and the German automobile association. Through working groups, the committee developed five main project proposals:

- neighbourhood-based service centre in public-private partnership;
- a "mobility agency" offering tailor-made advice on the reduction of trips and alternative transport arrangements to institutions, companies or individuals;
- an advisory and lobbying agency for housing co-operatives, focussing on the reduction of planning and financing costs;
- establishment of an annual "Agenda 21 Award" for an exemplary project of sustainable urban development practice in the Munich region;
- "co-operative planning practice" - guidelines for early and effective co-operation between planners and citizens, initially focussing on the "Theresienhöhe" urban development project.

Results

In spring 1998, the consultation phase of the LA 21 process drew to a close, with proposals for a sustainable Munich being presented to city council in the form of a comprehensive report. Therefore, the process of creating an LA 21 document reached at least a preliminary conclusion. It seems important to distinguish between the material outcomes of those four years and the process results that may be of greater importance for urban development in Munich in the long run.

However, the transformation of ideas into action was only then beginning, and with it arose the need for a reformed organizational structure, as the focus shifted from establishing a widespread consultation process to detailed project work. In this sense, any analysis of results can be only tentative, and we will return to the changes of the post-consultation phase a little later.

Material results. First and foremost among the material results of the LA 21 process in Munich is the existence of a formal LA 21 document. This in itself is no small achievement, as it is based on an extensive and participatory consultation process. The document contains guidelines for actions and decision-making by council as well as proposals for specific projects.

These projects form the most tangible results of the LA 21 process. Altogether, 44 such projects have been developed. Not surprisingly, given the diversity of issues covered by the special committees, they vary greatly in themes and approaches. Some projects cover "classic" issues, while others break rather new ground, especially when it comes to combining the different dimensions of sustainability. There is also a notable variety of envisaged outcomes: some projects aim at producing information material (e.g., guide books, videos, reports), some seek to set up advisory services (e.g., on integrated environmental management for private enterprises, ecological housing construction and renewable energy use), others aim at direct and specific action (e.g., setting up ethical investment funds) or even new institutions (e.g., the above-mentioned neighbourhood service centre or a "one-world-centre"). Almost all these projects are jointly carried by several initiatives, enterprises and municipal departments, often bringing together actors that had few if any contacts in the past.

What has proven problematic, however, is the organization and financing of project work, which means that the projects vary greatly in terms of state of realization. The special committee on "housing, settlement and mobility" illustrates this problem fairly well, as virtually all its projects developed serious problems in securing finance or even adequate manpower. While concepts were drawn up rather successfully during the consultation phase, their realization has been largely lacking for various reasons. These range from problems of communication to a lack of institutional support by the planning department, who is apparently unwilling to see its own agendas challenged.

Particularly noteworthy on the positive side, however, is that some activities of consultation and communication have been recognized as being projects in their own right. The summer campaign of 1997, which tried to increase involvement of the public in LA 21 discussions, is one such project, the establishment of an open forum discussing the Theresienhöhe urban development project is another one. And even the whole LA 21 process itself is seen as such a project. It aims at the development of a new culture of public dialogue characterized by the mutual respect of all actors involved and, at the same time, at providing opportunities for learning the social and communication skills such a dialogue requires.

Process results. LA 21 planning in and for Munich has therefore yielded process-oriented results that go beyond the tangible outcomes of individual projects. The work of the special committees and the advisory board involved some 200 "expert" citizens over the course of two years. The adult-education programmes run by MVHS were visited by some 2,000 individuals each semester, and the summer campaign of 1997 involved thousands more at neighbourhood level (e.g., through 40 neighbourhood festivals). Even though this is clearly not the 1970s' style of participation in urban planning, the last four years have seen a concerted effort to animate the general public to get involved in debating the future of Munich and thereby setting an agenda for politicians and other actors to follow. However, the exact benefits of this effort - in terms of increasing social cohesion and trust - are intangible and very difficult to evaluate.

More than anything, though, this has been a learning experience for all involved. The process has established new communication links and networks between the most diverse actors, and it has exposed many of them to new forms of dialogue and new techniques of communication. This point is repeatedly and independently stressed by most participants, even those who tend towards a very critical and pessimistic assessment. In various fields of urban development there has arisen a new atmosphere of exchange and co-operation among previously unconnected private actors. Similarly, public-private dialogue in those fields has reached a different quality, with private interests often pressuring civic administration to behave in a more co-operative way. In this sense, the LA 21 process as a whole has been an experiment in city-wide communication, devising forums and techniques capable of establishing and sustaining it. However, an overall assessment of process results must remain ambiguous, since the forces of inertia - especially in civic administration – remain strong.

Reorganization. With the large-scale consultation phase officially ended, the LA 21 process has entered a period of reorganization so that – almost unavoidably – a relative lull in acitivities and enthusiasm has set in. With the LA 21 document complete, the advisory board and special committees have fulfilled their role and work has supposedly shifted to the projects themselves. The agenda office has been reduced to two staff, both from the newly amalgamated department of health and environmental protection, while the planning department has withdrawn its staff. The agenda office maintains its co-ordinating role and faces the task of constructing a list of indicators to allow ongoing evaluation. A kind of enlarged steering committee has emerged called Network Future Munich to oversee the further process and to support co-ordination and information flows between the various actors still involved. And lastly, in late 1998 a "Citizens' Foundation" has been founded by private actors and initiatives to raise funds and to increase the financial contributions of the private sector in order to achieve a sustainable economic base for further activities. The city contributed DM 400,000 to the foundation in 1998 and promised up to DM 400,000 more in matching funds for 1999.

Figure 8.4 Revised structure of LA 21 in Munich

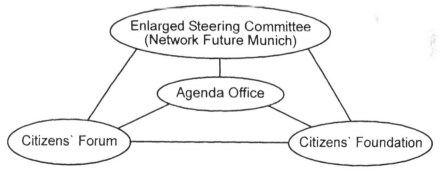

Preliminary Assessment. This lull of activity, caused partly by the need for reorganization, reveals some of the failures and frustrations of the Munich experience. Even during the consultation phase, some special committees were less productive than others, and some managed the integration of diverse interests better, some worse. More often than not, relative success seems to be related to the approach and attitudes adopted by the respective department in civic administration.

This is particularly notable for the planning department which used to be connected to the Special Committee on Housing, Settlement and Mobility. Since the end of the consultation phase, the department has abandoned the special committee, taking no steps to support and maintain the communication networks that had emerged. Most observers reflect upon the department's seeming unwillingness to alter its general routines for how urban development strategies are devised and planning processes are implemented.[13] In this sense, the impact of LA 21 on the department's routines of participation and urban development planning has been slight. The special committee's project ideas about new participatory approaches in the Theresienhöhe project were frustrated, and private initiatives sought out new projects where the power of the city's planning department is more constrained on account of the involvement of the state government.[14] On the positive side, there is an increasing tendency for private initiatives and citizen groups to demand more participatory involvement in planning and to make overt reference to LA 21 to justify these demands.

Hence, it remains to be seen whether the LA 21 process will produce a lasting change of urban development practice in Munich. The continuity of the process, the quality of debate and the follow-up of the political-administrative system are still in no way secured. Much will hinge on the sustainability of finance; the citizens' foundation seems a step in the right direction, but an independent future – without the city's contributions – remains uncertain.

Conclusions

Quality of life and participation, while long-established issues in urban development, have taken on a new meaning and importance in the 1990s. Participation in the 1970s started with citizen protest generally directed against modernization projects perceived as threatening people's individual quality of life. Urban social movements often combined with middle-class intellectuals and more environmentalist concerns to demand a say in urban development issues. Activists united in their resistance against what "those poli-

---

[13]   As mentioned before, the planning department has withdrawn its staff from the agenda office and declared different priorities: rather traditional urban development projects are viewed as more important than the participatory and ecological aspects contained in the LA 21 idea – though the department's official line (and rhetoric) would probably deny this claim.

[14]   The rivalry between the (conservative) state and the (social democrat) city government may partly explain the apparent resistance of the planning department. The state government has heavily emphasized sustainability and Agenda 21 issues in its policies, arguing that quality of life will become the prime economic development issue in the future.

ticians and planners" had in mind. Participation, in other words, emerged from a clash of values. The political-administrative system reacted to these demands by formalization: participation was recognized as a legitimate and necessary step in planning processes, and procedures were developed to integrate participation into the standard planning process.

Conditions for urban development in the 1990s are quite different. Post-industrial societies have become more differentiated socially, and economic restructuring has led to severe problems of unemployment while arguably leaving governments less powerful and more often in fiscal crisis. Most urban governments have reacted by using growth-oriented development policies. However, despite pressing economic problems, a new environmentalism has emerged in line with a much stronger, refined focus on community values. The novelty of the situation lies in the integrated perspective of these formerly separate or conflicting views: maintaining a healthy environment and a stable social community is now seen as one of the keystones of a successful economic strategy – particularly at the local level. Hence, cities are scenes of social conflict (and disintegration) and of investment decisions increasingly based on quality of life (and environmental) issues. The concept of "sustainability" illustrates this integrated perspective, and the sustainability debate has shifted notably to the local level, making "sustainable urban development" arguably the key topic for research and urban development planning.

Through this process, urban development has also changed. It tends towards a more cooperative consensus-oriented approach. The way ahead is seen as a concerted partnership effort with diverse views not necessarily a hindrance but as potentially beneficial. Partnerships may mobilize hitherto unused resources – social, personal and financial. On the one hand, this has shifted the focus from abstract programmes towards specific projects, since tangible concerns can mobilize more actors. On the other hand, urban development is starting to be seen as a continuous and open-ended discourse. Rather than spending energy in confrontation and protest, citizens are now becoming motivated to actively participate in debating and shaping futures.

The Munich case illustrates several of these tendencies. The LA 21 process is more open and more flexible, it has involved new actors and resources and led to new networks of communication. Nevertheless, it would be unrealistic to announce a fundamental shift in urban development. Flexibility of approach does not mean that new procedures, techniques and styles are replacing "old" ones, but that their use is carefully and consciously orchestrated when deemed appropriate or expedient (cf. Selle, 1994). The resistance of some key actors illustrates the considerable forces of inertia in the political-administrative system. However, the pressure is clearly on, and this manifest resistance is probably the best evidence that practices are indeed changing and that other actors are forcefully shaping a new stage for local urban development. Whether the necessary level of enthusiasm and activity will prove sustainable in itself remains an open question.

Let us conclude with two cautionary notes, both stressing the political dimension of the process. First, the basic premise of the LA 21 approach, implicit both in UN declarations and in the documents and activities at the local level in Munich, is that wider participation will ultimately lead to "better" urban development and thus to greater quality of life for urban citizens.[15] As yet this remains a premise, its empirical validity – even for Munich – is still virtually impossible to assess. To state that more participation is implicitly good (as more democratic) simply leads to a circular argument. And second, important issues of democratic practice still need to be addressed, e.g., the uneven distribution of power among actors or how to deal with problems too controversial for consensus. Foremost among these seem to be questions of legitimacy and control, for the less regulated and more participatory approach of LA 21 could conceivably strengthen more corporatist tendencies, directly challenging established systems of representative parliamentary democracy at the local level.

---

[15]   Cf. the idea of empowerment as explicated e.g., by Friedmann (1992).

Window: who is a participant?

Very active forms of citizens participation, beyond the game of the political parties, have long been observed in the cities. In the nineteenth century, some residents movements were active, fighting for the improvement of the neighborhood. The last thirty years has seen a strong renewal of such demands, with the development of the new social movements partly constituted on urban issues. The social composition of these two types of participation, of course is different: in the first case, the local bourgeoisie, established in the neighborhood for a relatively long time, was the most active, while in the second it was new social categories, characterized more by their cultural than economic capital.

Nowadays, as the electoral participation is weakening, there is even more interest in examining other forms of participation. Some important results can be mentioned in this context.

Even if there are big differences in the rate of associative participation in the European countries, higher in the Nordic ones, lower in the Mediterranean ones, there is no global decrease these last twenty years: associative participation is still important.

The social profile is more or less the same as that of electoral participation: overrepresentation of higher educated people, under representation of foreigners and lower classes.

Even if the social profile is similar, it is not the same inhabitants who participate in the political and associative spheres. In this sense, associative participation represents an alternative to other forms. The result is that associative participation represents a form of civic engagement that is important for the legitimacy of the urban authorities as well as for the building of local citizenship.

This form of participation is nevertheless far from being perfect and raises various questions: the definition of who can legitimately participate is not formalized in any institutional rule. Some specific mobilizations of one or another group can change the result of the consultation. During the process, some important actors can change position, regardless of their previous involvement. Finally, the use of proximity to mobilize people is not without danger: the immediate interest of a neighborhood could sometimes contradicts a policy beneficial to the city as a whole. Nevertheless, more and more local governments are conscious of the interest to increase direct information and, sometimes, even citizen participation through the associative movements.

Dominique Joye

References

ALBERS, G. (1988) *Stadtplanung: Eine praxisorientierte Einführung* (Die Geographie). Darmstadt: Wissenschaftliche Buchgesellschaft.

BANNER, G. and REICHARD, C. (Eds.) (1993) *Kommunale Managementkonzepte in Europa: Anregungen für die deutsche Reformdiskussion.* Köln: Deutscher Gemeindeverlag.

BASTEN, L. (1998) *Die Neue Mitte Oberhausen: Ein Grossprojekt der Stadtentwicklung im Spannungsfeld von Politik und Planung* (Stadtforschung aktuell, 67). Basel, Boston and Berlin: Birkhäuser.

BUNDESMINISTERIUM FÜR RAUMORDNUNG, BAUWESEN UND STÄDTEBAU (Eds.) (1996) *Lokale Agenda 21* (Schriftenreihe "Forschung", 499). Bonn: Bundesministerium für Raumordnung, Bauwesen und Städtebau.

BUNDESMINISTERIUM FÜR UMWELT, NATURSCHUTZ UND REAKTORSICHERHEIT (Eds.) (n.d.) *Konferenz der Vereinten Nationen für Umwelt und Entwicklung im Juni 1992 in Rio de Janeiro: Dokumente – Agenda 21.* Bonn: Bundesministerium für Umwelt, Naturschutz und Reaktorsicherheit.

CAMPBELL, S. and FAINSTEIN, S.S. (Eds.) (1996) *Readings in Planning Theory.* Cambridge (USA) and Oxford (UK): Blackwell.

DANGSCHAT, J.S. (1996) Zur Armutsentwicklung in deutschen Städten, in: AKADEMIE FÜR RAUMFORSCHUNG UND LANDESPLANUNG (Eds) *Agglomerationsräume in Deutschland: Ansichten, Einsichten, Aussichten,* pp. 51-76. Hannover: Akademie für Raumforschung und Landesplanung.

DAVID, B. and BOCK, S. (1998) Eine Stadt kommt in Schwung, in: ÖKOM – GESELLSCHAFT FÜR ÖKOLOGISCHE KOMMUNIKATION mbH (Eds.) *Zukunftsfähiges München: Ein gemeinsames Projekt Münchner Bürgerinnen und Bürger,* p. 47-50. München: Ökom.

DIENEL, P.C. (1978) *Die Planungszelle: Der Bürger plant seine Umwelt. Eine Alternative zur Establishment-Demokratie.* Opladen: Westdeutscher Verlag.

DRESCHER, B.U. and DELLWIG, M. (1996) *Rathaus ohne Ämter: Verwaltungsreform, Public-Private-Partnership und das Projekt Neue Mitte in Oberhausen.* Frankfurt/Main and New York: Campus.

DUNCAN, S. and GOODWIN, M. (1988) *The Local State and Uneven Development: Behind the Local Government Crisis.* Cambridge: Polity Press.

ECKARDT, W. (1997) Die Münchner Agenda 21 oder das Märchen vom süßen Brei, *das forum,* 37, pp. 20-26.

FRIEDMANN, J. (1987) *Planning in the Public Domain: From Knowledge to Action.* Princeton: Princeton University Press.

FRIEDMANN, J. (1992) *Empowerment: The Politics of Alternative Development.* Cambridge (USA) and Oxford (UK): Blackwell.

HALL, P. (1988) *Cities of Tomorrow: An Intellectual History of Urban Planning and Design in the Twentieth Century.* Oxford (UK) and Cambridge (USA): Blackwell.

HARVEY, D. (1989) From Managerialism to Entrepreneurialism: the Transformation in Urban Governance in Late Capitalism, *Geografiska Annaler B,* 71, pp. 3-17.

HAUGHTON, G. and HUNTER, C. (1994) *Sustainable Cities* (Regional Policy and Development, 7). London (UK) and Bristol (USA): Kingsley.

HEALEY, P. (1992) Planning Through Debate: The Communicative Turn in Planning Theory, *Town Planning Review,* 63, pp. 143-162.

HEINZ, W. (1993) Public Private Partnership: ein neuer Weg zur Stadtentwicklung?, in: W. HEINZ (Ed.) *Public Private Partnership - ein neuer Weg zur Stadtentwicklung?* (Schriften des Deutschen Instituts für Urbanistik, 87), pp. 29-61. Stuttgart, Berlin and Köln: Kohlhammer a.o.

HELBRECHT, I. (1994) *"Stadtmarketing": Konturen einer kommunikativen Stadtentwicklungspolitik* (Stadtforschung aktuell, 44). Basel, Boston and Berlin: Birkhäuser.

KRÄTKE, S. and SCHMOLL, F. (1987) Der lokale Staat: "Ausführungsorgan" oder "Gegenmacht"? *Prokla 68*, 17, pp. 30-72.

KREUZER, K. (1998a) Unterwegs zur Nachhaltigkeit, in: ÖKOM – GESELLSCHAFT FÜR ÖKOLOGISCHE KOMMUNIKATION MBH (Eds) *Zukunftsfähiges München: Ein gemeinsames Projekt Münchner Bürgerinnen und Bürger*, pp. 23-26. München: Ökom.

KREUZER, K. (1998b) Auf dem Weg zu einem zukunftsfähigen München, in: INTERNATIONALER RAT FÜR KOMMUNALE UMWELTINITIATIVEN; S. KUHN, G. SUCHY and M. ZIMMERMANN (Eds.) *Lokale Agenda 21 – Deutschland: Kommunale Strategien für eine zukunftsbeständige Entwicklung*, pp. 193-202. Berlin a.o.: Springer.

LASH, H. (1977) *Planning in a Human Way: Personal Reflections on the Regional Planning Experience in Greater Vancouver* (Urban Prospects). Ottawa: Ministry of State for Urban Affairs.

LORIMER, J. (1978) *The Developers*. Toronto: James Lorimer.

LÖTSCHER, L. (1985) *Lebensqualität kanadischer Städte: Ein Beitrag zur Diskussion von methodischer und empirischer Erfassung lebensräumlicher Qualität* (Basler Beiträge zur Geographie, 33). Basel: Geographisches Institut der Universität Basel, Geographisch-Ethnologische Gesellschaft Basel.

LÖTSCHER, L. and KÜHMICHEL, K. (1998) Lokale Agenda 21: partizipative Planung nachhaltiger Stadtentwicklung? *Geographica Helvetica*, 53, pp. 135-138.

MAYER, M. (1991) "Postfordismus" und "lokaler Staat", in: H. HEINELT and H. WOLLMANN (Eds.) *Brennpunkt Stadt: Stadtpolitik und lokale Politikforschung in den 80er und 90er Jahren* (Stadtforschung aktuell, 31), pp. 31-51. Basel, Boston and Berlin: Birkhäuser.

NEWMANN, P. and THORNLEY, A. (1996) *Urban Planning in Europe: International Competition, National Systems und Planning Projects*. London and New York: Routledge.

ÖKOM – GESELLSCHAFT FÜR ÖKOLOGISCHE KOMMUNIKATION MBH (Eds.) (1998) *Zukunftsfähiges München: Ein gemeinsames Projekt Münchner Bürgerinnen und Bürger*. München: Ökom.

SELLE, K. (1994) *Was ist bloß mit der Planung los? Erkundungen auf dem Weg zum kooperativen Handeln. Ein Werkbuch* (Dortmunder Beiträge zur Raumplanung, 69). Dortmund: IRPUD.

SELMAN, P. (1996) *Local Sustainability: Managing and Planning Ecologically Sound Places*. London: Chapman.

UMWELTSCHUTZREFERAT DER LANDESHAUPTSTADT MÜNCHEN (Eds.) (1997) *Münchner Agenda 21, Zwischenbericht 1997*. München: Umweltschutzreferat der Landeshauptstadt München.

WALLACE, I. (1992) *The Global Economic System*. London: Routledge.

WORLD COMMISSION ON ENVIRONMENT AND DEVELOPMENT (Eds.) (1987) *Our Common Future*. Oxford: Oxford University Press.

ZIMMERMANN, M. (1997) Lokale Agenda 21: Ein kommunaler Aktionsplan für die zukunftsbeständige Entwicklung der Kommune im 21. Jahrhundert, *Aus Politik und Zeitgeschichte*, B 27/97, pp. 25-38.

Part III
Urban Governance
and Planning

9 New Towns and Compact Cities: Urban Planning in The Netherlands between State and Market

WIM OSTENDORF

Introduction

Answering the question "What has and what has not changed in the urban system of The Netherlands?" one has to address the role of urban planning in The Netherlands. The question is even more relevant in view of the changes taking place in urban planning in recent years, causing people to wonder if urban planning remains a major structural force in the formation of Dutch cities and in creating satisfying living environments in The Netherlands.

This important question emerges within the framework of a globalizing economy, increasing international competition (Sassen, 1991) and, with the collapse of the communist world, a retreat from state intervention creating room for market-conformity. This question certainly applies to The Netherlands, where "national physical planning is quite unique in having fifty years of cumulative experience" (Wusten & Faludi, 1992, p 18) and where a considerable shift in the ideas related to physical planning has been observed based on changing international relations.

However, for several reasons the question is not very easy to answer. First, we have to limit our observations to The Netherlands, leaving open for discussion the extent to which the Dutch experience can be generalized to other countries. Second, the question has to be treated more precisely. It suggests that urban planning has been very influential in the past, but has lost significance in recent years. Although this sounds plausible for the case of The

Netherlands, a solid answer to the question requires more objectivity. Third, the problem arises as to what is meant by the influence of urban planning. Is it just the number of professionals working in this field, is it the measurement of a result that would not have emerged up without the impact of urban planning, or is it a question of the success of urban planning in creating situations considered superior to the situations that would have appeared without urban planning. The question reflects interest in the results of planning. Nevertheless, planning is difficult to evaluate. The system of physical planning is not very helpful in this respect. Hence, we must first pay attention to the criteria for judgement (section 2). Next, we will try to evaluate the now completed new town policy in Dutch physical planning (section 3) and the new compact city policy (section 4). We will then try to apply these findings to a broader framework (section 5) in order to answer the question mentioned above (section 6).

Evaluating planning: plan conformity and goal conformity

A basic assumption of this paper is that the future role of urban planning depends on the degree of success of present planning policies. Of course, the role of planning may also depend on more ideological factors: planning as a perceived necessity irrespective of the results. In general however, a market orientation dominates the present ideological climate: everything has to be justified by its success. We do not expect this situation to change very fast. This brings us to the measurement of the success of physical planning. This is not easy, either. Planners, like others, are not very eager to discuss their own legitimacy by evaluating their planning. Moreover, goals are not very clearly defined, making evaluation even more difficult (see window). Thus, our first problem is finding criteria for the judgment of planning.

Here 'plan conformity' and 'goal conformity' are used as criteria. 'Plan conformity' refers to the extent to which the projects have been carried out according to the plans, while 'goal conformity' refers to the achievement of the goals of planning by judging the effects or results. Using these two approaches, we will try to evaluate whether the new town policy in The Netherlands been successful? This results in the following questions:

- Has the new town policy in The Netherlands been carried out according to the plans?
- Did the new town policy meet its goals?

- And we will try to evaluate whether the compact city policy in The Netherlands will be succesful, resulting in the following questions:
- Will the compact city policy be carried out according to the plans?
- Will the compact city policy meet its goals?

The success of the new town policy

After the Second World War The Netherlands had to cope with large population growth and an even greater shortage of housing. This was not the only point of concern, however. It was realized that population density in the western part of the country, the economic core region, was already very high, and overcrowding was feared. As a matter of fact, planners held a anti-urban attitude. Dutch cities, they asserted, should not have more than one million inhabitants, and that is what certainly would happen if the new houses were to be built in the cities of Amsterdam, Rotterdam and The Hague. Moreover, due to considerable growth, the population was expected to increase by 7.5 million inhabitants to 20 million in the year 2000. In case of non-intervention, a suburban sprawl would cause the cities and villages in the western part of the country to grow together, resulting in one amorphous megalopolis and decreasing the quality of life (Van Engelsdorp Gastelaars & Ostendorf, 1991). This 'doom scenario' perspective (Van der Wusten & Faludi, 1992, pp. 22-23) formed the background for the ideas in the Second Report on Physical Planning (Ministry of Housing and Physical Planning, 1966): the policy of no concentration of population and no urban sprawl would be achieved by 'bundled (or concentrated) deconcentration'. On first sight this presents a paradox: a struggle against urbanization and suburbanization at the same time, that had to be realized by establishing economically independent new towns. Part of the policy involved preserving open space between cities and especially keeping the agricultural area in the middle of the Randstad, the Green Heart, free from urbanization (figure 1). Thus, the new towns (or 'growth towns' since most were not completely new) had to be built in an outward direction, some at considerable distances from the donor city, fitting in with the idea of a more equal distribution of the population across the country. The construction of suburban housing in new towns, affordable by a large share of social housing, was completely in accordance with the housing preferences of Dutch working - and middle class families, for whom suburbanization was an expensive ideal.

Figure 9.1 The Randstad, the Green Heart and the new towns

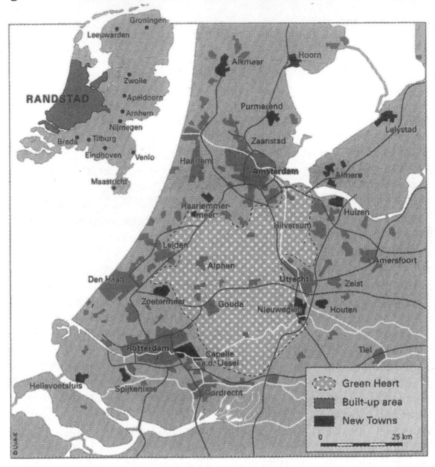

Summarizing these ideas, we can state that the premises of the new town policy were to stop urban growth, to prevent suburban sprawl, especially in the Green Heart, and to construct new towns. The goal was to improve the quality of life of the inhabitants in general and their housing situation in particular, in the cities, but even more so in the new towns.

The new town policy concentrated on preventing the growth of settlements other than the designated new towns and on keeping the Green Heart free from suburbanization. During the 1970s this did not yet operate very effectively: suburban growth was everywhere. Moreover, the government proved unable to improve the employment situation in new towns, because companies and firms were not very interested in the peripheral location of the new towns and government had no instruments to force them. Therefore, the new town policy concentrated on housing. In this field the government held powerful financial instruments by providing partial or full financing for the housing. So, in the 1980s, with a strong recession on the housing market, the plan conformity of the new town policy as far as house building was concerned, became very high. Faludi (1994, p 493) is impressed by these results: "most (new towns) exceeded their targets, and there are now more than half a million witnesses to the success of this policy, i.e., those people who have migrated to where successive government documents said they ought to go". Such plan conformity, Faludi suggests, would make even the former communist planners in Eastern Europe jealous.

The relocation of jobs to the new towns, however, remained a problem, especially for the more peripheral new towns. The result was a weak employment situation (Ministerie VROM, 1986; Van Engelsdorp Gastelaars & Ostendorf, 1991). Thus, it was very difficult to find a job in such a new town, resulting in high commuting distances (especially among men), massive traffic jams and high unemployment rates (especially among married women). In addition, the declining situation in the cities due to the loss of population to suburbs and new towns also caused problems. The decrease in the number of inhabitants, of the more affluent in the first place, resulted in a reduction of the local average income. Moreover, the economic crisis of the late 1970s and early 1980s hit especially the large cities, resulting in economic decline and increasing unemployment. Economic growth did not take place in the big cities, but in centrally located suburbs and medium sized towns. Conformity to the goal of improving the quality of life was thus much more difficult to achieve than conformity to the new town plans.

The success of the compact city policy

The compact city policy has to be understood as a reaction to the new town policy which had to be corrected: new towns were functioning as dormitory towns, there was urban decline, increasing commuting distances and increasing mobility by private car leading to environmental problems. The compact city approach was also a reaction to increasing international competition following the collapse of Eastern Europe and in an integrating European Community.

Table 9.1   Possession of private cars per square kilometre and
             per 1.000 inhabitants in some European countries,
             1996

Country	number of private cars per square kilometre	number of private cars per 1,000 inhabitants
The Netherlands	141	371
Belgium	142	427
Austria	44	461
Germany	115	500
Italy	105	554
Luxembourg	90	558

Source : Eurostat

The Fourth Report of Physical Planning (Ministerie VROM, 1988) and the Supplement (Ministerie VROM, 1990) concentrate on economic aspects - international competition, the role of the free market, government withdrawal - and on environmental aspects, aiming at spatial configurations, where all functions (housing, employment, facilities) can be found at short distances and, as a consequence, where mobility is low. Reducing the growth of individual mobility is considered imperative, because The Netherlands is unique in Europe in terms of traffic-density: a very high car-density per square kilometer, while private ownership of cars is still relatively low (table 1). Thus, while further growth in mobility is to be expected, the capacity of the highways is at its maximum with no possibility for further extension. This is why the new town policy had to come to an end. Revitalization of cities became the new policy goal. Moreover, the four big Dutch cities were assigned the role of 'motors' of the Dutch economy in order to generate economic growth for the Netherlands.

The compact city concept sought to end the population loss of cities in favour of suburban locations. Plans to realize the creation of the compact city include: new locations for houses and for employment to be found in or adjacent to existing cities in the vicinity of public transport facilities, and 'balancing' the urban population by diminishing the proportion of low-income 'social' housing. The concept also involves a shift to a more market-oriented approach, which means that the proportion of social housing is to be reduced and people should pay the full amount of their housing costs. House-building in suburban locations and constructing new offices near highways not accessible by public transport, especially in the Green Heart, was to be prevented by this policy. Hence, the solution to environmental problems (creation of the compact city) should result in economic prosperity in the cities and in the country as a whole. To put it briefly, the plans call for the creation of compact cities; the goals are to reduce (the growth of) individual mobility, protect the environment and to create prosperity.

Figure 9.2   Increase of the housing stock in Amsterdam 1970-1993

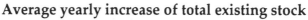

Average yearly increase of total existing stock

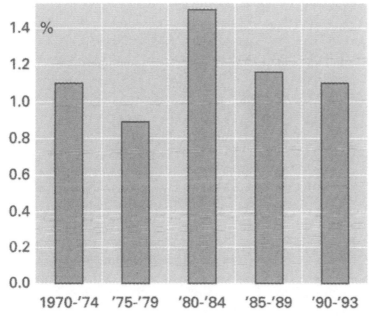

Source: CBS

The compact city policy has not yet operated for a sufficiently long time in order to be able to provide a solid judgement as to its plan conformity or goal conformity. Nevertheless, serious doubts can be raised. These doubts are not seen in the opinion of the National Physical Planning Agency, whose interim evaluation states that: "There is no need to correct the policy of the Fourth Report on Physical Planning and of the Supplement. The developments that have led to the Reports did not lose anything of their relevance. Now the plans have to be put into effect. ... There is no need for a New Report on Physical Planning" (Ministerie VROM, 1994; also Van Staalduine & Drexhage, 1995). In general, one can say that the need for the compact city is much more clear that its feasibility. Hence, let us try to assess the feasibility of the compact city policy.

Figure 9.3   Increase of the housing stock as a percentage of total existing stock 1986-1993

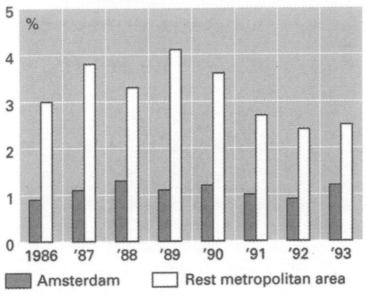

Source: CBS

Related to plan conformity, a positive result is the cessation of construction of houses in new towns. This is primarily caused by the withdrawal of governmental money. However, government is unable to pay the costs for the creating the compact city. For the most part, residential construction projects must be based on a market-based approach and pay for their own completion and development. This makes participation of the market indispensable. In general however, one can say that the private sector is not very much interested in the compact city. Owner occupiers are much more interested in buying a house in a suburban environment than in the compact city (Wassenberg et al., 1994). The compact city policy, however, contradicts the preference of Dutch family households for suburban housing. Hence up to now, there has hardly been any growth in the urban population; and if so, this growth is caused by foreign immigration and not by any movement back to the city.

Figure 9.4   Tenure structure, social rental as a percentage of total number of newly built housing 1986-1993

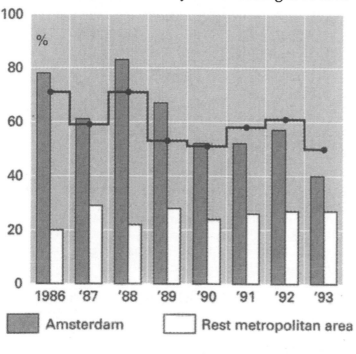

Amsterdam

Rest metropolitan area

—•— Amsterdam/rest area [1])

Source: CBS
[1]) absolute number of social housing built in Amsterdam/
absolute number of social housing built in rest
metropolitan area

The suburban orientation of the population did not change when the government switched to a compact city policy. As figures 9.2, 9.3 and 9.4 show, the compact city policy did not yet result in any change in the construction of new houses in Amsterdam and its urban region: Amsterdam appeared unable to fundamentally alter its annual increase in the housing stock (figure 9.2); moreover, the increase remains considerably higher outside the city, in the remaining part of the urban region (figure 9.3). Given these shares of new construction, Amsterdam hardly succeeded in changing the tenure structure of the new houses: otherwise, as called for by the plans, the new construction of Amsterdam remains very much dependent on subsidized or social housing (figure 9.4).

Table 9.2   Mean income and location of households in the urban
             regions of Amsterdam, Rotterdam and The Hague in
             1981 and 1989/1990, by quartiles

Year		1981				1989/1990		
Quartile	1st	2nd	3rd	4th	1st	2nd	3rd	4th
cities	28	28	24	20	30	27	23	20
new	13	17	31	40	17	21	28	34
towns								
suburbs	20	21	26	32	16	22	28	34
Total	25	25	25	25	25	25	25	25

Source: Central Bureau of Statistics, Housing Demand Surveys 1981, 1989/91.

Table 2 reveals some doubts about the urban revitalization
that the compact city policy is intended to bring about: the changes
in the mean income of the population of the different parts of the
urban regions of Amsterdam, Rotterdam and The Hague between
1981 and 1989 fail to show the desired increase in the three cities; as
could be expected, the new towns are losing position, but unlike
the intention of the policy, the position of the suburbs improved
even more.

Doubts on urban revitalization brought about by the compact
city policy are even more increased by table 9.3, which brings the
period of the early 1990s into the picture. Table 9.3 shows the difference in average income between each of the four largest Dutch
cities and their respective suburban environments. It shows that
the differences that increased during the 1970s were rather constant
during the 1980s and increased again in the 1990s. These changes
mirror the waves of suburbanisation and the relative absence of
this process during the economic crisis of the 1980s, but may also
be connected to a general rise of income inequality in the Netherlands since 1987 (SCP, 1996). Hence, during the 1990s, cities are
again losing ground with respect to relatively wealthy residents. In
the Netherlands, it is feared that the urban population is becoming
poorer and state dependent (SCP, 1996). Table 9.33 shows that the
compact city policy did not result in the desired improvement in
the position of the four cities during the early 1990s.

The halt to housing construction in the new towns, together
with the very limited shift of construction to the cities, has resulted
in a considerable shortage of dwellings. Moreover, it appears very
difficult to find large, cheap locations in or adjacent to the cities,
where new houses can be built; most available locations appear to
be small and expensive. As cited above, the Government considers
this as a problem related to the start of the new policy, while others
see it as a sign of the failure of the policy.

Table 9.3   Difference between the average income per earner
(total income) of the central city and the urban region
(minus central city) relative to the urban region, 1974,
1984, 1989, 1994

	1974	1984	1989	1994
Amsterdam	15.8	16.7	15.7	18.8
Rotterfam	10.1	15.9	15.5	18.2
The Hague	10.3	11.7	11.2	11.7
Utrecht	13.2	17.2	18.9	21.0

Source: Statistics Netherlands.

Firms did not lose their suburban orientation either. In general, the market shows a much larger interest in locations near highways, accessible by private cars than in locations in the vicinity of public transport.

Related to the compact city policy, some form of regional government is considered to be very helpful in a succesful international competition of Dutch cities. In referenda, however, the urban populations of Amsterdam and Rotterdam showed no enthusiasm for losing their identity as the inhabitants of these cities. In general, then, one can say that the government wants the compact city; cooperation of the market, however, is indispensable, but absent. This will result in the failure to realize the compact city policy.

When the plans will not be realized, it seems hardly possible to achieve the goals. Moreover, one can state that it seems very difficult to combine the goals of the compact city policy: revitalization of the Dutch cities, economic growth and a reduction in individual mobility. Related to employment, there is a fear that if one cannot find a suitable location and/or if the use of the private car proves to be difficult or impossible, firms and businesses may migrate to other parts of the country or even to other countries. Thus, a too rigid emphasis on concentration and public transport might result in the migration of people and jobs which does not bring the goal of revitalization of cities or economic growth of the country any closer. This tension clearly appears in the case of Amsterdam Airport (Schiphol) and the port of Rotterdam. Growth of these two mainports and a good infrastructure for both is considered indispensable for future economic growth in The Netherlands. Here the government seems to be the prisoner of two goals that are not easily combined: improvement of international competition in good accessibility versus environmental protection in limitation of mobility.

Summarizing, in my view the compact city policy will not be successful; it will fail to realize the plans, and it will not achieve its goals.

A typology of policy-situations

With the help of table 4, we now try to bring the findings with respect to the difference between the success of the new town policy and the compact city policy into a broader framework. Table 4 contains a typology of policy-situations, characterized by two dimensions: who is paying for the investments in the physical planning, government or market; and whose interests are predominantly served by the physical planning-policy, individual or general interests.

Table 9.4   A typology of policy-situations

		Individual interests	General interests
Investments in physical planning are paid by	Government	1	2
	Private sector	3	4

This typology shows two 'balanced' situations: Situation 2, in which the government invests for the sake of general interests; and Situation 3, in which actors on the market invest for their individual interests. Situations 2 and 3 are balanced with respect to desirability and feasibility. In Situation 1, considerations of feasibilty seem dominant: the government is investing for the sake of individual interests. It seems to be impossible for such a policy to fail. The opposite is found in Situation 4: considerations of desirability dominate considerations of feasibility: the market has to act for the sake of general interests, and this seems to be like rowing against the current.

This typology points to obvious conclusions. The construction of houses in the new town policy indicates Situation 1: the government invested in the construction of houses, which were strongly desired by the individual households. This policy can be labeled as 'accomodating'. Therefore, the plans could (easily) be realized, making the policy successful. The government did not invest in the creation of jobs in the new towns. Firms did not seriously consider relocating to these outlying new towns, because this was not in the interest of the firm but a general interest; the characteristics of Situation 4. As a consequence, this part of the plans of the new town policy failed.

The compact city policy is also characterized by Situation 4: the market must act for the sake of general interests. This policy does not try to be accommodating but is trying to prevent situations that the government considers being undesirable. Considerations of desirability dominate considerations of feasibility. The experience with the relocation of jobs in the new town policy justifies the expectation of failure. This can be called the intrinsic weakness of this present policy.

Of course, this leads to the question how this can be changed. The typology suggests adaptation in two directions: (1) in the direction of Situation 3, by giving more attention to individual interests; in other words, to the demandside; (2) in the direction of Situation 2, with a stronger role for the government. This last adaptation seems far from the present ideological climate; for the basic assumption is that the market must do the job.

Conclusions

According to Van der Wusten & Faludi (1992, pp. 18-19) "Dutch national physical planning is quite unique in having fifty years of cumulative experience. So as a function of government, planning is well established, with a National Physical Planning Agency advising a minister responsible for physical planning ... Since the 1970s the meaning of planning has been in dispute, and physical planning has lost some of its appeal ... So the present is a time of uncertainty for planners." These somewhat sad words are in accordance with our conclusion that the new towns have been realized according to plan, but that the policy did not reach its goals. The same conclusion was reached by Aldridge for the British new towns, as already appears from the very clear title of the book: "The British New Towns: A Programme without a Policy." (1979). The change to the compact city policy underlines the same conclusion.

The evaluation of the compact city policy is even worse: neither plan conformity nor goal conformity will be achieved. For the realization of plan conformity, the government needs the support of the private sector, but the private sector is not very interested in the compact city. The reason for this negative attitude is that the creation of the compact city is not considered in the interest of the market partners. Goal conformity is difficult to achieve if the plans cannot be achieved. Moreover, even if the plans of the compact city will be achieved, it is very doubtful if this will result in the revitalization of the Dutch cities and in economic growth for the country.

All the evidence points in the direction of more economic growth in the suburban parts of the urban region.

One can formulate the same conclusion in a different way. The new town policy, as well as the compact city policy, are compromises, in which a terrifying scenario had to be avoided with the help of a very attractive justification. In order to avoid an amorphous megalopolis, new towns had to be built; the construction of new towns was justified by housing preferences (the population was in favour of the new towns) and succeeded because they were built with government money. However, the migration of employment proved to be much more difficult than the migration of population. The resulting situation appeared to be much less desirable afterwards because of the dormitory character of the new towns, the increasing commuting distances and urban decline.

In order to protect the environment and to halt urban decline, compact cities have to be constructed. The compact city is justified by the promise of economic growth, but it will not succeed because marketforces, which are supposed to pay for the realization of the plans, are not very much interested: neither employment nor population is in favour of the compact city.

This judgement brings us back to the question of the changes in the urban system of The Netherlands and to urban planning as a major structural force in the formation of cities and in creating satisfying living environments. The new town policy has resulted in new towns as a new urban reality in The Netherlands; the compact city policy will be less influential. The question about the present influence of urban planning can hardly be doubted: "is physical planning still applicable?" If the perspective is directed at The Global City, as sketched in the work of Sassen (1991), little room seems left for planning. This perspective seems to be the background of the question. Yet if the answer is based on a prediction of its results, as carried out above, the answer is not very positive. However, practice can be different. In many cases the need to interfere is operating more than the perspective of success! For instance, the perspective of the growth of mobility and the insufficient capacity of the Dutch infrastructure is much stronger than the idea of achieving a reduction, even when success is absent. Next to this, the idea of planning can change from comprehensive to just one or some sectors, e.g. infrastructure. Future planning has to be more modest, with a very careful equilibrium between goals and instruments. In general we must conclude that it remains difficult to predict the future of physical planning.

The evaluation of planning

Over the last decades studies evaluating the rate of success of government policies have become quite widespread. This also applies to the field of planning, where evaluations ask whether the specific policy brought new or simply predictable results. It seems quite obvious to pose such questions. It is less clear how to investigate them. This problem is caused by several factors. In the first place, the goals of policies are often not formulated in a clear and specific way. Second, insofar as such policies are formulated, their objectives are rather vague and not easily translated into clear, objectively measurements.

Faludi & Van der Valk (1990) mention two approaches to the measurement of success: implementation research looking for "plan conformity" and evaluation research looking for "goal conformity". Plan conformity refers to the extent, to which the projects have been carried out according to the plans, while goal conformity refers to the achievement of the goals of planning by judging the effects or results. "There is a difference between the two approaches: implementation research, for instance, may demonstrate that a project has been carried out in conformity with the plan, while evaluation research leads to the conclusion that the goals of the plan have not been met at all!" (Dieleman & Van Engelsdorp Gastelaars, 1992; p 76). Hence, the difference between the two can be summarised in the following cynical statement: "operation successful, patient dead".

Implementation research assessing "plan conformity" is more easy to carry out than evaluation research searching for "goal conformity". This is caused by the fact that the plan is far more clearly formulated than the goals of a policy. Politicians tend to be prudent in formulating clear goals, as this may result in a dangerous tool for judging whether the politician is right or wrong; vague notions always offer a smart politician a way out. This did not hold true for the "read my lips"-statement of president Bill Clinton, promising no increase in taxes; this statement became famous for the very reason that it gave his opponents a clear means of proving that the president was wrong. To avoid such embarrassment, clear goals are seldom formulated. For the same reason, policy documents do not contain clear sets of goals to be achieved. Because of this, the researcher must assume the task of formulating the goals of the policy, based on the problems that the policy intends to tackle. In such cases, the research must start with a very precise study of relevant policy-documents.

Another problem of evaluation research involves the measurement of abstract goals: how do we judge an improvement in the quality of life or an increase in social mobility? Implementation research is less subject to this kind of problem; plans concerning the number of houses to be built or renovated are much easier to measure.

Since implementation research is easier to carry out than evaluation research, it is also more popular among researchers. However, evaluation research is more fundamental and thus more needed.

Wim Ostendorf

References

ALDRIDGE, M. (1979) *The British New Towns: A Programme without a Policy*. London.

CORTIE, C., M. DIJST and W. OSTENDORF (1992) The Randstad a Metropolis? *Tijdschrift voor Economische en Sociale Geografie*, Vol. 83, No. 4, pp. 278-288.

DIELEMAN, F.M. and R. Van ENGELSDORP GASTELAARS (1992) Housing and physical planning, in: F.M. Dieleman and S. Musterd (Eds.) *The Randstad: A Research and Policy Laboratory*, pp. 65-96. Dordrecht The Netherlands: Kluwer.

FALUDI, A. (1994) Coalition building and planning for Dutch growth management: the role of the Randstad concept, *Urban Studies*, Vol. 31 No. 3, pp. 485-507.

KREUKELS, A. (1992) The restructuring and growth of the Randstad cities - current policy issues, in: F.M. Dieleman and S. Musterd (Eds.) *The Randstad: A Research and Policy Laboratory*, pp. 237-262. Dordrecht The Netherlands: Kluwer.

Ministerie Vrom (1986) *De Toekomst van de Groeikernen. Een verkennende Studie*. Den Haag: Rijksplanologische Dienst.

Ministerie Vrom (1988) *Fourth Report on Physical Planning*. Den Haag: Staatsuitgeverij.

Ministerie Vrom (1990) *Fourth Report on Physical Planning - Supplement*. Den Haag: Staatsuitgeverij.

Ministerie Vrom (1994) Balans van de Vierde Nota Ruimtelijke Ordening (Extra). Ruimtelijke Verkenningen 1994. Den Haag: Staatsuitgeverij.

Ministry of Housing and Physical Planning (1966) *Second Report on Physical Planning*. Den Haag: Staatsuitgeverij.

OSTENDORF, W. (1988) *Het Sociaal Profiel van de Gemeente*, Ph.D.Thesis, Amsterdam: Instituut voor Sociale Geografie.

PRIEMUS, H. (1994) Planning the Randstad: between economic growth and sustainability, in: *Urban Studies*, Vol. 31 No. 3, pp. 509-534.

SASSEN, S. (1991) *The Global City*. New York, London, Tokyo. Princeton: Princeton University Press.

SCP (1996) *Sociaal en Cultureel Rapport 1996*. Rijswijk: Sociaal en Cultureel Planbureau.

SHACHAR, A. (1994) Randstad Holland: a "world city"?, in: *Urban Studies*, Vol. 31 No. 3, pp. 381-400.

VAN DER WUSTEN, H. and A. FALUDI (1992), The Randstad - playground of physical planners, in: F.M. DIELEMAN and S. MUSTERD (Eds.) *The Randstad: A Research and Policy Laboratory*, pp. 17-38. Dordrecht The Netherlands: Kluwer.

VAN ENGELSDORP GASTELAART, R. and W. OSTENDORF (1991) New towns: the beginning and end of a new urban reality in The Netherlands, in: M.J. BANNON, L.S. BOURNE and R. SINCLAIR (Eds.) *Urbanization and Urban Development. Recent Trends in a Global Context*, pp. 240-249. Dublin: University College.

VAN STAALDUINE, J. and B. DREXHAGE (1995) The state of the nation after five years of national spatial planning, *Tijdschrift voor Economische en Sociale Geografie*, Vol. 86 No. 2, pp. 191-196.

WASSENBERG, F. et al. (1994) *Woonwensen en realisatie van VINEX-locaties in de Randstad*. Den Haag: Ministerie van VROM.

10 Urban Change, Planning and the Organization of Everyday Life. The Case of Örebro

ANN-CATHRINE ÅQUIST

Introduction

Cities change perpetually and in various respects. The changes can be analyzed in different ways. The present article concerns changes in the Swedish city of Örebro during the last fifty years. The focus will be on changes in urban planning, especially in housing. One important change which has occurred in this period concerns the main problems or themes that are focused upon in planning. Fifty years ago the main problem was housing provision: there was a shortage of housing and many dwellings were in bad condition. This problem has since been remedied. Today an important theme in planning is environmental issues. I have chosen to study the changes in urban planning in a certain perspective, that of everyday life.

The main part of the present article consists of an analysis of the master plan for Örebro from 1955. The focus is on the representation of everyday life. One idea behind studying the representation of everyday life in the 1955 plan is to get a reference-point which can be used for addressing the issue of how to plan for environmentally friendly organizations of everyday life. Since environmental problems have become an important political issue during the 1990s, urban planning must take into consideration how an environmentally friendly everyday life can be supported through planning.

After this introduction follows a presentation of the city of Örebro and then a discussion of the perspective of everyday life. After that comes the presentation of the analysis of the master plan for Örebro from 1955. At the end of the article planning for an environmentally friendly everyday life is discussed.

The city of Örebro and the local economy

Örebro is located in the south of Sweden, about 200 kilometers west of Stockholm. The municipality of Örebro, including the city's hinterland, has about 120 000 inhabitants. It is the sixth-biggest city in Sweden. The economy is fairly diversified, with a large tertiary sector. Örebro is a regional administrative center, and it has a university and a regional hospital.

From the beginning of the 20th century until 1970 the city experienced a yearly growth in population due to in-migration. The in-migration was especially large from 1945 to 1970. During the 1960s the city's population increased by app.roximately 1600 - 1800 each year (Egerö 1979, p 35). In 1970 the increase in population was reversed into a decrease. The city lost population during most years of the 1970s. The migration flows turned around again in 1980 and since then in-migration has been larger than out-migration, though not as large as in the 1960s. Between 1980 and the mid-1990s the population growth was about 400-500 a year (Municipality of Örebro 1995).

The changes in population in Örebro are connected to changes in the local economy. From the beginning of the 20th century until the 1960s the city's economy was dominated by manufacturing industry, especially in footwear, textiles, and engineering. The industrial restructuring which started in the 1960s hit the local economy of Örebro hard. By the early 1970s almost all manufacturing companies in textiles and footwear had been closed down. Also engineering industry decreased but the consequences were less dramatic. Unemployment and social problems followed in the wake of the economic restructuring. In the same period the service sector expanded in Örebro. In the 1960s the fast growth of the three largest urban areas in Sweden - Stockholm, Gothenburg and Malmö - was generally considered a problem. It caused overcrowding, housing shortage, increased prices, etc, in the three metropolitan areas and problems connected to the out-migration in the countryside and smaller towns, especially in the north of Sweden. The parliament passed a number of resolutions designed to prevent further metropolitan growth through decentralizing various activities. Subject to these activities were cities functioning as regional centers in the urban system, mainly capitals of the counties. Specialized health care is one. Higher education was established in Örebro as well as in a few other cities where it had not been established before. Part of the Swedish Bureau of Statistics was moved from Stockholm to Örebro. During the 1960s and the 1970s the base of the local economy of Örebro changed, almost as a textbook-example, from manufacturing industries to service industries.

Provision of housing

One main issue in planning and politics in Örebro from the 1940s and to the 1970s was the housing problem. The large in-migration to the city led to a big housing shortage. In the late 1960s app.roximately 18 000 persons were registered as being in the line for housing (Egerö 1979, p 35). Many houses were in bad condition and lacked facilities like hot water, bathroom and indoor toilet. The housing problem was not unique to Örebro but prevailed in all Swedish cities at the time. Housing policy was an important part of the building of the Swedish welfare state, which started when the Social Democratic Party won the parliamentary election of 1932. In the 1940s the Social Democratic government encouraged the municipalities to set up municipal housing companies. These companies were given special state-subsidized mortgages and the commission to come to grips with the housing shortage. This implied a new involvement in the housing problem on the part of the state as well as of the municipalities.

The Swedish housing policy from the late 1940s until 1965 is characterized as experimental (SOU 1981:100). One successful experiment was the housing estate Rosta in Örebro. It was built by the municipal housing company of Örebro (ÖBO) in the period 1947 to 1952. It has a special physical form: the three-story houses are in the form of three-point stars which are connected in long chains, billowing through a large park area. Rosta was built with 1343 apartments, most of which had two rooms and a kitchen, and 30 percent of which had three or more rooms (Egerö 1979, p 133). In the 1950s these were not considered small apartments. The basic planning ideal for Rosta was the neighborhood unit. It had a service center with food stores, a branch of the public library, a day-care center and schools.

After Rosta, ÖBO built several housing estates. The two earliest, Rosta and Baronbackarna, are even today considered successful ones. This gave Örebro the label "model-city". Here Social Democratic housing policy became realized to a larger degree than in many other cities. The municipal housing company in Örebro played an unusually important role in housing provision (Strömberg 1989).

In 1965 the Social Democratic government launced a program for intensifying housing contruction, since the housing shortage was still severe. The aim of the program was to build 100 000 apartments per year for ten years. Now housing estates became larger, monotonous, and less varied in design. Often the design of the apartments was of high quality and the apartments were well-equipped, but the estates as such were too large and too monotonous. By 1973 the housing shortage seemed to have been remedied. Empty apartments began to appear, and in the fall of 1973 ÖBO had 100 empty apartments. This was, at least partly, caused by the recession in the economy in the early 1970s, which stopped the migration flows to the city. Around 1970 the policy of large housing estates met with criticism for giving rise to bad living conditions. The demand for single-family houses and row houses increased. From 1975 housing policy shifted in favor of this type of dwellings.

Everyday life and its conditions

Everyday life can be defined as the activities or practices through which we reproduce ourselves and thereby indirectly reproduce society (Bloch 1991). One characteristic strand of everyday life is its routinization.

> "... when the day begins, its programme is already printed and the stage is set for another performance; much like yesterday's."
> (Parkes and Thrift 1980, p 218)

We usually do the same things day after day, though the routines of week days often differ from routines of weekends. Both paid work and unpaid work are important factors with regrad to how everyday life is organized. Responsibilities and the care of others, children and elderly persons, also have a strong impact on everyday life.

Through daily routines everyday life acquires an organization in time. There is also an organization in a longer time perspective, in the variations between the different phases of life. For a small child everyday life looks different from what it does to a teenager or a middle-aged person. There is a spatial perspective as well

(Bohm 1990). Our everyday life is situated in social space. Different activities are located in different places. Different groups of people use to some extent different spaces. The social segregation of the city is an example of this.

A distinction can be made between the everyday life and its conditions (Bloch 1991). The conditions are the social and societal preconditions for the everyday life. This includes paid work, housing, education, health care and other services, but also cultural norms like ideas of "the good life". Conditions include as well the organization of everyday life in the perspective of time and space. Here urban planning is important since it shapes the spatial organization of the city. The design of housing, the relative location of various activities in the city, the transportation system are all factors which have an impact on the conditions for people's everyday life. For example, the planning of the transportation system in a city is crucial with regard to the possibility of choosing to bike or travel by public transport to work. Another example of the importance of urban planning for the organization of everyday life is the location of food stores and child-care centers in a city. In a concrete situation the conditions of everyday life can facilitate or hinder activities.

Henri Lefebvre on planning and everyday life

The French philosopher Henri Lefebvre has written about the city and everyday life. He argues that in every city plan there is a concealed program for everyday life (Franzén, Sandstedt 1982). A plan for the city is in some way based on ideas about the kind of everyday life that will be lived in the area covered by the plan. These ideas are usually not explicitly stated but are taken for granted. They will shape the conditions for everyday life. Lefebvre's idea that the program for everyday life is concealed can be interpreted in different ways. One interpretation is that each period of time has its own general and taken for granted understanding of how people's everyday life is organized. This interpretation is reflected in urban planning as well as in other spheres, without being discussed or questioned. Another interpretation is that the main theme of a plan can be for example transportation - anything but the organization of everyday life. However, this will have consequences for the organization of everyday life, even if these consequences are not investigated. Though Lefebvre's idea can be interpreted in various ways, nevertheless it opens up interesting questions about planning from the perspective of everyday life. In this perspecitve I will make an analysis of the master plan for Örebro from 1955.

The master plan for Örebro 1955

One reason to choose a plan for the city of Örebro is that its urban planning and housing design from the 1940s and onwards has earned Örebro the epithet of model city. Örebro has been considered a good example of successful Social-Democratic housing policy. In the introduction to Bertil Egerö's book (1979) on housing policy and construction in Örebro, Sven Thiberg writes:

> "Örebro has been a model city for housing politicians and planners since the 1940s. Far-sighted land policy, openness towards new planning ideas and the power to carry through big projects brought the city into the limelight."
> (Thiberg in Egerö 1979, p 6)

The first master plan for Örebro was presented in 1955. It was made by the city planning architect Herman Hermansson. It consists of about 80 pages of text and several maps. It covers the time up to 1970. It is based on a population forecast from 1947, made by William William-Olsson, professor of geography. This forecast predicts the population of 1970 remarkably well; it also predicts a stagnation in population growth in the 1970s. Reading between the lines of the plan you can find a representation of everyday life. The master plan concerns several issues that shape the conditions for the everyday life of the population. Here, plans are made for the spatial organization of the city: transportation, localization of work-places, housing, various types of services like schools, day care centers, post offices, sports grounds, churches.

Transportation in the city

The master plan contains an idea of the physical form of the city's built-up area. It consists of the city center and five "bands" coming out of the center in the shape of a star. Partly the city already had this form, with built-up areas along the river which runs through the city, and along the main roads out of the city. The reason for designing the built-up area of the city in the shape of a star is to make transportation within the city as convenient as possible. In the middle of each "band" runs an arterial street to the city center, like a spoke of a wheel. These streets are designed to have public transportation. The distance from a dwelling to a bus stop should be no more than 400 meters. Theoretically this constrains the breadth of the "band" to 800 meters, according to the plan. The importance of transportation in the plan is also illustrated by the reasoning about some areas where the land consists of clay soil. This land is not well-suited for the construction of multi-family houses. Instead of departing from the idea of a star-shaped built-up area the plan argues for a technical solution (piles) that will make it possible to build on the land with clay soil. It is argued that in the long run this will be cheaper than having a less efficient itinerary for public transportation.

The modes of transportation which are considered within the city are mainly biking and public transportation (buses). Private cars were not part of everyday life in Sweden in the 1950s. In 1953 there were 60 cars per 1 000 inhabitants in Sweden, whilst the corresponding figure for the U.S. was 284 (master plan for Örebro 1955, p 55). The plan contains a discussion of the future of car-ownership in Sweden. Car-ownership is assumed to be going to increase, though probably without reaching the U.S. figure. There are several arguments for this in the plan. The physical layout of Swedish cities is much more compact. It would be difficult to accommodate more cars in the city centers without destroying valuable historical buildings, and economically it would be unrealizable. In the plan it is said that in the future more people will probably get cars but use them mainly at weekends.

> "Most of the cars that the inhabitants of Örebro will get in the future will probably be used mainly during leisure time, for trips to the countryside, to bathing places and summer houses on Saturdays and Sundays."
> (my translation) (master plan for Örebro 1955, p 58)

In the 1950s the private car was mainly used for leisure activities. It was not part of the organization of everyday life as it is today.

Neighborhood units

The "bands" which come out of the city center are designed mainly for housing but also for a few industrial areas. The industrial areas are located adjacent to the housing areas. The two should not be mingled - here is a zoning idea. The housing areas should be designed as neighborhood units, a planning idea developed by the American planner Clarence Perry in the 1920s (Hall 1988). Perry argued for the design of housing estates as units with dwellings and frequently used services, like food stores and schools. A neighborhood unit should have the size of the catchment area of an elementary school. The school was to be a community school, i.e. functioning as a space for various meetings, after school hours, for people in the neighborhood. Separation of different modes of transportation was part of Perry's idea. The area should have pedestrian streets and the school should be reachable on foot. Cars were only allowed on the main streets surrounding the neighborhood unit.

The master plan for Örebro states that housing should be constructed in neighborhood units, where dwellings are located around a center with schools and food stores. People should be able to do their daily shopp.ing, on foot, within the neighborhood unit. The plan does not include the idea to separate different modes of transport, but it is stated that the distance to a school for the youngest children should not be more than 500 meters and for older children no more than 1 000 meters. It is assumed that a neighborhood unit needs between 2 000 and 6 000 inhabitants in order to supp.ort a sevice center.

Dwellings in multi-family houses

The buildings in the neighborhood units will be multi-family houses. When the plan was made about 80% of all dwellings in Örebro were of this type. For alleviation of the housing shortage and bad housing conditions it was supp.osed that the multi-family house was the most efficient. The master plan refers to a study, made in 1946, about housing conditions and people's preferences. According to the study, more than 40% wanted to live in a single-family house. This is not taken seriously by the planners. It is said that people have no experience of modern dwellings in new multi-family houses and do not know the problems of living in a single-family house. The master plan states that two thirds of the dwellings to be built in the 15 years covered by the plan shall be constructed in multi-family houses and the rest in single-family houses, some in the form of row houses.

The localization of housing in relation to workplaces

In the master plan, the star-shaped form of the city's built-up area is discussed in connection with the localization of housing areas in relation to work-place areas. The dominating work-place area is the city center, but also some other industrial areas contain many work-places - at that time many factories, for example the footwear factories, were in the city center. It is stated that people should have the opp.ortunity to live close to their work-place. This was not the case when the plan was made; there was "an exceptionally large traffic of cyclists" on the two bridges over the river in the very center of the city when the work-day started and ended (master plan for Örebro 1955, p 19). The situation is concidered to be explained by the shortage of (modern) housing. As soon as the shortage is remedied, it will be possible for people to move to an apartment close to their work-place. This reasoning presupposes that every dwelling is related to *one* work-place, i.e. only one person in every household is assumed to have a job (outside the home). If two persons in a houshold have jobs then the localization of the home in relation to work-places becomes more complicated. It is said in the plan that supp.osedly the husband's work-place will be decisive when a family looks for housing. But in the future, industrial areas should be planned so that both male and female workers are demanded in the same area.

Women's paid work and the demand for daycare centers

In the 1950s most married women in Sweden were full-time housewives. The plan refers to statistics, from 1945, showing that in Örebro only 10.8% of the married women were gainfully employed. However, the statistics only include fulltime employment despite the fact that, even then, many married women had parttime employment. According to a study based on income statistics instead of employment statistics, about half of the married women with an income had a fulltime job in 1950 (Nyberg 1985). If parttime employment is included, then app.roximately 22% of the married women in Örebro were gainfully employed in 1950. After WW II Swedish manufacturing industry experienced a shortage of labor. Married women were seen as a potential resource labor, even though the housewife ideal dominated. In the master plan this is illustrated by the following example:

> "The question of the future supp.ly of labor for manufacturing industry is very difficult to assess. We know very little about the extent to which women will be rid of housework and join the labor market, neither do we know the possibilities of an increase of parttime jobs in various branches, or the consequences of this."
> (my translation) (master plan for Örebro 1955, p 27)

Women's labor is assumed to be tied to the home and the tasks connected to caring for a home and a family, at least at present. Though the future might bring changes. The plan contains a chapter on daycare centers and pre-schools. The demand for daycare centers is said to be difficult to predict since fertility rates vary and even more important is "married women's opportunities and wishes to be gainfully employed" (ibid, p 43). In 1955 there were 5 daycare centers in Örebro which corresponds to one daycare center for 14 000 inhabitants. According to the plan, this covered the demand at the time, but it is assumed that in the future the demand will increase. So, a reservation of land for 11 daycare centers is made.

A program for everyday life

The type of everyday life that is represented in the master plan is the life of a family with a bread-winning father, a home-making mother and children of school age. They live in a multi-family house in a housing area which has been planned as a neighborhood unit. The father goes to work, in the city center or in the adjacent industrial area, by bike or by bus. He probably comes home for lunch in the middle of the day. The mother can do most of her errands, on foot, in the small service center of the neighborhood. For some of her errands she needs to go to the city center, then the bus is a convenient mode of transportation. The children can walk to the nearby school, and the younger children stay at home with their mother. If the family has a car they use it to go out of town on weekends.

It is hardly surprising that a plan from 1955 is based on this image of people's everyday life. The family consisting of a bread-winning father, a home-making mother, and children was an ideal for many people, men and women, at the time. In Sweden, the 1940s and 1950s is sometimes labelled "the era of the housewife". In that period a majority of the married women were full-time housewives. Earlier, as well as later, married women have to a larger extent worked outside the home. (Hirdman 1983, Åquist 1987) However, the question of how well this type of everyday life suited all people might very well be raised.

Efficiency

In the master plan for Örebro from 1955, there is a focus on efficiency in transportation, mainly public transportation. There is also a focus on housing, since it was important to alleviate the bad housing conditions and the housing shortage. The plan contains an idea of the organization of people's everyday life, and how this organization can be improved by planning. The physical arrangement of various activities in the city makes it possible for families to have a certain efficiency in their everyday life. That kind of efficiency can be promoted by, for instance, good public transportation and by the idea of the neighborhood unit, as it was perceived in Sweden with emphasis on various services in housing areas. In the plan the representation of everyday life is used as a tool to plan for efficiency.

Important changes since 1955

There have been important changes in demographic factors since the 1950s, when the first master plan was made. Immigration has increased. In the 1950s there were hardly any immigrants in Sweden. In the mid-1990s more than half a million of the inhabitants of the country are not Swedish citizens (Swedish Bureau of Statistics 1996).[1] Women's participation in the labor market has increased. The full-time housewife has disapp.eard, both as an ideal and as a reality. Today, 80 % of Swedish women are gainfully employed, 45 % have fulltime work (Swedish Bureau of Statistics 1996). Fertility rates are lower now than in the 1950s. In 1995 the total fertility rate was 1.7 (Swedish Bureau of Statistics 1996). Household composition has changed. Today, 35 % of households consist of only one person. Only 19 % of households consist of a couple with children.

Another important change has to do with the growth in car-ownership. This has had an important impact on the physical organization of the city. To the car-owners the area within everyday reach has increased a lot, which has led to changes in the organization of everyday life. It is now possible to have various services and recreational facilities spread out. The functional separation of activities is more far-reaching. This has also led to increased differences between those who have access to a car and those who do not.

There have also been changes in housing. In the 1950s many people lived in apartments in multi-family houses. Today it is more common to live in a single-family house, a detached house or a semi-detached row house. Many people who work in the city of Örebro live outside the city and commute to their work-place.

---

[1] In 1994 Sweden had 8.8 million inhabitants (Swedish Bureau of Statistics, 1996).

Commuting has increased very much during the last three decades, which of course is connected to the growth in car-ownership.

There have been considerable changes in the focus of local politics since the 1950s. In the 1950s and the 1960s housing provision and urban renewal were main issues in the municipality of Örebro as well as in most other (urban) municipalities in Sweden. In the 1980s economic problems in the municipalities became important. There were cuts in budgets, and efficiency problems in municipal service provision and administration were discussed. In the 1990s, environmental problems and sustainability have app.eared as a new area of interest of local politics and urban planning in Örebro, as well as in other Swedish municipalities.

Towards planning for sustainability

The importance of environmental problems and sustainability was manifested by the United Nation's Earth Summit in Rio de Janeiro in 1992 on environment and development (UNCED). Here sustainable development was presented as a basic principle. In Agenda 21, one of the programs of action adopted at the conference, local authorities (municipalities) are given important roles when it comes to realizing the goals of sustainability. (Blowers 1993, Department of Environment and Natural Resources, 1993) One of the tools at hand for municipalities is land-use planning. According to Bob Evans:

> "During the next decade, the profession and practice of land-use planning in Britain will need to change and adapt to meet the demands of a new century and a growing concern with environmental issues." (Evans 1997, p 13)

In the master plan for Örebro from 1955, the representation of everyday life is used to plan for the provision of housing. Today, environmental issues can be said to have replaced the provision of housing as a leading theme of urban planning. I would suggest that representations of everyday life can be used as a tool for planning for an environmentally friendly city, just as it earlier was used as a tool for planning for the provision of housing. This raises the important question of the content of an environmentally friendly life. The suggestion that a representation of an environmentally friendly life can be used as a tool for planning for sustainability must be accompanied by an idea of what characterizes such organization of everyday life.

When trying to formulate what characterizes an environmentally friendly life it is appropriate to distinguish between urban life and country life, since preconditions vary considerably between cities and the countryside. In the following, only urban life will be discussed.

The National Federation of Tenants' Associations in Sweden has published a small book on environmental issues in connection with housing, called "The Tenants' Book on the Environment". It was published in 1997. The dwelling and its surroundings are important preconditions for the organization of everyday life. From what is stated in the book about housing and the environment an idea of what constitutes the environmentally friendly life can be constructed. This does not mean the there is only *one* form of everyday life that is environmentally friendly. Probably, several forms of everyday life can be considered environmentally friendly, depending on circumstances such as what is dominant in the debate about environmental issues at the time.

From "The Tenants' Book on the Environment" the preconditions in the dwelling and its surroundings for an environmentally friendly life can be constructed in the following way:

- The opp.ortunity to buy environmentally friendly products like detergents and other cleaning products, low-energy light bulbs, and ecologically produced food.
- Equipment in the house which facilitates an environmentally friendly everyday life: water-saving dishwashers, washing machines, faucets and toilets, measuring instruments for water and energy consumption for every household, storage space for sorted waste, equipment and space for composts, equipment for sorting of sewage water.
- The environmentally friendly house shall have clean air, app.ropriate temperature and air humidity. It shall be free from emissions from building materials, from radon radiation, from damp and mildew, and from noise from neighbors, fans, traffic, etc.
- The environmentally friendly house shall be located in a suitable micro-climate concerning: solar radiation, vegetation, wind directions, cold air pockets and radon in the bed-rock. The air shall have as low a degree as possible of pollution.
- The immediate surroundings of a house shall have a garden for play and recreation or a park in the vicinity, allotments, trees which subdue noise and to some extent clean the air, as well as protect from winds, etc.
- In the vicinity of the house there shall be stations for collection of sorted waste.

To this list, constructed from "The Tenants' Book on the Environment", environmentally friendly transportation can be added. An important part of an environmentally friendly everyday life is to have access to public transportation and bicycle lanes. To this car pools can be added. Another thing which has a bearing on transportation in an environmental perspective is the market for locally produced food. This form of selling food is environmentally friendly since it does not require long transport.

The list illustrates that different actors and organizations are important in order to create conditions for environmentally friendly living. Also the choices and actions of the individual are important, for example when it comes to choosing among different brands of detergents. But this too, requires a context where such choices are possible - there must be environmentally friendly detergents on the market. The equipment of the house is important. Certain equipment can make it easier for the tenant to save water and electricity. Composting of organic household waste requires specific equipment. The individual tenant's everyday agency cannot influence whether the sewage water is separated or not, that depends on whether the drains of the house, the drain system of the city, and the municipal sewage treatment works are arranged for separation or not. The same applies to the provision of electricity. The individual cannot choose to use energy from an environmentally friendly source but must accept what comes in the electricity wire. The individual has only a limited opp.ortunity to influence the micro-climate in the place where the house is situated, as well as the surroundings. The individual can choose to move to a place or a building with good environmental conditions, but that presupposes that urban planning and building contractors have done their part. To summarize, the individual has certain possibilities of organizing everyday life in an environmentally friendly way, but is also dependent to a large extent on whether possibilities are provided or not by urban planning and other municipal activities, as well as on the markets for housing and various products (Åquist 1998). This illustrates that urban planning has an important role to play in solving environmental problems.

Conclusions

Societal change brings changes in the central themes of planning. In the 1950s Swedish urban planning was focused on housing problems, which included both housing shortage and bad housing conditions. Transportation in the city was also an important issue.

In the 1990s the societal situation is different. New problems have appeared which bring new themes to urban planning. For example, some of the large housing areas, built in the 1960s and the 1970s, have become "problem areas" - that is, areas where various social problems have become concentrated and thereby aggravated. This is an important issue for urban planning, especially in the metropolitan areas of Stockholm, Gothenburg and Malmö. More generally, environmental problems have become main themes of urban planning in the 1990s. Since the 1950s the focus of urban planning can be said to have changed from housing provision to environmental isues.

In urban planning, there can be various points of departure for handling the main theme of a plan. One example is the structure plan for Gothenburg from 1992, where environmental issues are central. Here, one point of departure in the planning process is eco-cycles of water, nitrogen, and energy (Municipality of Gothenburg 1992). Another point of departure for the planning process is the organization of everyday life. According to Henri Lefebvre, a representation of how everyday life is, or could be, organized is always there in urban planning. Every urban plan contains a program for everyday life, whether the planners are aware of it or not. In the master plan for Örebro from 1955 there is a fairly explicit idea of the organization of one form of everyday life: of the nuclear family with a breadwinning father, a housemaking mother, and a few children, living in an apartment in a housing area planned as a neighborhood unit. In this plan, the organization of everyday life can be said to be used as a planning tool, as a way to handle the main theme of the plan. The same can be done in planning for sustainability. Now, when environmental problems have become such important issues in planning, representations of the environmentally friendly everyday life can be used as a planning tool.

From authoritarian to participatory ideals in planning

During the 1950s, when the master plan for Örebro was made, the Swedish urban planning system was authoritarian but also benevolent. It was authoritarian in the sense that planning was considered a task for experts, guided in their work by scientific knowledge. Science and experts were supp.osed to have the knowledge required in order to find the best solutions to planning problems. They formulated the needs of the citizens. The planning system was also authoritarian in the sense that it was highly centralized. The state issued directions for planning in the municipalities. There were quotas for the number of dwellings that every municipality was allowed to build. If a plan should gain legal status it had to be passed by state authorities.

Despite the authoritarian character of this planning system, it was also in a sense benevolent. One motivation for the planning system was the bad housing conditions and the shortage of housing. Remedying this was an important part of the building of the Swedish welfare state. There were strong egalitarian traits. For example, the municipal housing companies, founded in the late 1940s, constructed housing, of high quality, which was aimed for everybody. This was not "social housing", as for example in Britain, where state-financed housing was open only to certain groups of the population.

Today, the Swedish planning system looks different. The tasks are not the same as in the 1950s. The financial situation is much more problematic. The welfare state has to some degree been dismantled. These changes have been accompanied by a decentralization process, where the central state has transferred more and more tasks to the municipalities and state regulations have diminished. In the Planning and Building Act, from 1987, urban planning is handed over to the municipalities. The Act emphasizes the importance of participation of citizens in the planning process, a participation which still has not found its forms. This indicates a different way of thinking about the knowledge required for planning, whereby also citizens are seen as having knowledge and experience to contribute. Another change is the increased diversity of the Swedish population. Immigration and the emergence of various lifestyles have made the population much more diverse. It is no longer app.ropriate to assume that planning should be designed for one generic type of family.

Ann-Cathrine Åquist

References

BLOCH, C. (1991) I lust och nöd - om vardagsliv och känslor. (For better or for worse - on everyday life and emotions.) *Kvinnovetenskaplig tidskrift* nr 2 1991, pp. 31-42.

BLOWERS, A. (1993) Environmental Policy: The Quest for Sustainable Development, *Urban Studies* Vol. 30, No. 4/5, pp. 775-796.

BOHM, K. (1990) Teorier om vardag och struktur (Theories of the everyday and structures), in: *Fysisk planlegging i forvandling: natur - struktur - hverdagsliv*. Rapport fra Nordplans 20-årssymposium Stockholm: Nordplan, pp. 73-106.

DEPARTMENT OF ENVIRONMENT AND NATURAL RESSOURCE (Miljö- och naturresursdepartementet) (1993) *Agenda 21 - en sammanfattning*. (Agenda 21 - a summary) UNCED-biblioteket, Stockholm.

EGERÖ, B. (1979) *En mönsterstad granskas. Bostadsplanering i Örebro 1945-75*. (An Examination of the Model-City. Planning for Housing in Örebro 1945-75) Stockholm: Byggforskningsrådet.

EVANS, B. (1997) From town planning to environmental planning, in ANDREW BLOWERS and Bob EVANS (eds.) *Town Planning into the 21st Century*. London: Routledge.

FRANZEN, M. and SANDSTEDT, E. (1982) Boendets planering och vardagslivets organisering - kvinnan, familjen och staden (Planning of Housing and the Organization of Everyday Life - Woman, Family and City), *Kvinnovetenskaplig tidskrift* No. 1 1982, pp. 6-15.

HALL, P. (1988) *Cities of Tomorrow*. Oxford: Blackwell.

HIRDMAN, Y. (1983) Den socialistiska hemmafrun (The socialistic housewife), in: B. ÅKERMAN et.al. (eds.): *Vi kan, vi behövs! - Kvinnorna går samman i egna föreningar*. (We can, we are needed! - Women form organizations of their own.) Stockholm: Akademilitteratur, pp. 11-59.

MUNICIPALITY OF GOTHENBURG (Göteborgs kommun) (1992)*Kretslopp. i översiktsplaneringen*. (Eco-cycles in structur planning.) Översiktsplan för Göteborg Planeringsunderlag 2:92.

MUNICIPALITY OF ÖREBRO (Örebro kommun) (1955) Generalplan för Örebro 1955 (Master plan for Örebro 1955).

MUNICIPALITY OF ÖREBRO (Örebro kommun) (1995) Örebro Statistik 1995:8 (Statistics from the municipality of Örebro).

NATIONAL FEDERATION OF TENANTS' ASSOCIATION (Hyresgästernas Riksförbund) (1997) *Hyresgästernas Miljöbok*. (The Tenants' Book on the Environment) Stockholm.

NYBERG, A. (1985) *Kvinnors arbete i industri och hem 1870-1910*. (Women's work in industry and at home 1870-1910) Tema-T arbetsnotat, Tema Teknik och social förändring, Linköpings universitet.

PARKES, D. N. and THRIFT, N. J. (1980) *Times, Spaces and Places*. London: Wiley.

SOU (1981) *Stadsförnyelse - kontinuitet, gemenskap, inflytande*. (Urban Renewal - Continuity, Community, Influence) Stockholm.

STRÖMBERG, T. (1989) *Mönsterstaden. Mark- och bostadspolitik i efterkrigs-tidens Örebro*. (The Model-City. Land and Housing Policy in Post-War Örebro) Center for Housing and Urban Research, University of Örebro.

SWEDISH BUREAU OF STATISTICS (1996) *Statistisk årsbok 96* (Statistical Yearbook of Sweden 96) Stockholm.

ÅQUIST, A-C. (1987) Hushållens tidsanvändning. En jämförelse mellan 1950 och 1983. (The use of time in households. A comparison between 1950 and 1983) Mimo. Department of Geography, University of Oslo.

ÅQUIST, A-C. (1998) Environmental Concern in Everyday Life. The Example of Introducing a New Waste-Handling System. *Scandinavian Housing & Planning Research* Vol. 15 No. 4, pp. 249-264.

11 Governing Functional Urban Regions - is the Principle of Subsidiarity Useful?

JOHN NOUSIAINEN

Introduction

There is a major need for an integrated institutional approach to the environmental and regulatory challenges experienced in the greater metropolitan areas of Europe.

Europe's post-war history contains many ambitious attempts to govern and manage large functional urban regions, thereby creating one overall metropolitan government with smaller ancillary administrative units e.g., the Greater London Council and the Openbaar Lichaam Rijnmond evolving around Rotterdam. All but a few have been abolished. This second wave of metropolitan governments has been provoked by change - change generated from both external and from internal sources. Increased integration among the European Union's member states and the globalisation of the economy have generated increased competition among urban regions. In this setting, various levels of governments are trying to improve welfare in urban areas and ensure balanced social and economic development. However, the spatial extension and interaction within the urban fabric, the development of new information and communication technologies, the crises of the welfare state - financial, but also intellectual in its inability to produce relevant policies with which to tackle the problems which have arisen - have once again focused on "good governance". Efforts are being directed towards achieving a consensus between the principal actors and the "common goal" (Lefévre 1998).

In this chapter, adequacy of scale, division of responsibilities and urban dynamics will be discussed. Focus will be on the principle of subsidiarity. Using a Danish example, I will demonstrate the difficulties in creating practical and workable new tiers of government.

Governance at the right level - but how?

To deal with changing economic, environmental and social circumstances, precise and efficient governance of urban regions is important. The principal questions is; "What administrative level represents the optimal level between effectiveness, efficiency and democracy?"

The debate is a normative one, implying that the different tasks and issues overlap perfectly, e.g., that those and only those tasks that by nature require local involvement should be claimed by the local level. In reality the division of tasks is likely to be much more complex, varying by issue and "emotional-investment" and hence not so clear-cut. No single or correct answer can be justified. Politicians prefer that decisions and responsibilities should be located as close to the citizens as possible. However, the problem is how to select which levels should be responsible for specific tasks and decisions. Some decisions should be made at the local level, others at a regional level, while still others must be taken at national or supranational levels. Smaller units with increased competence and power can be regarded as encouraging and very important for democracy as a whole. Yet overall objectives such as environmental protection and social welfare are at stake because the urban system as a whole may lose its power to regulate and plan for the benefit of the local or regional community if economic performance and spatial solidarity do not overlap thereby minimising problems such as NIMBY effects. An entity responsible for a certain policy and its implementation should, of course, also be entitled to procure and dispose of the financial, legal and other means to carry out such obligations (Jørgensen, 1996).

There is a need to co-ordinate strategic decision-making, not only for the city centres, but also for the functional urban region (FUR) as a whole, in order to solve the problems experienced. Adequacy of scale and proper size are not straightforward matters. Designing planning authorities which correspond to the boundaries of functional regions is a complicated affair that involves economic, administrative, social, cultural, environmental, transportation and ideological aspects, each with their own functional urban hinterlands. Trying to incorporate the different scales is a well-known geographical problem.

Admittedly, it is difficult to circumscribe urban space, to fix limits to the urban area, but nonetheless it remains true that there exists a territory which may be termed "functional".

However, the functional territory and the actual territory have different character. This raises a number of questions; How do we combine functional administrations which can deal with quantitative problems and solutions from a very rational point of view with legitimacy by the people who do not necessarily feel part or identify themselves with a larger geographical area (March & Olsen, 1989)? It is an extremely sensitive and delicate matter to conjoin the rational with the emotional (Larsen & Nousiainen, 1998). To what extent is it possible to combine the top-down approach (solving functional problems) with popular legitimacy (making sure that they feel appreciated, important and engaged)? The dynamic nature of the various tasks, functions and identity implies that scale and problems change over time and must therefore periodically be evaluated (Larsen and Nousiainen, 1998). Within the last decade, the principle of subsidiarity has been raised as a mechanism for determining the division of tasks and responsibilities. The following section will try to clarify this often unwieldy term.

The Principle of Subsidiarity - abstract goal rather than a method

> "It is by this partition of cares, descending in gradation from general to particular, that the mass of human affairs may be managed for the good and prosperity of all."
> Thomas Jefferson

In response to external and internal developments, numerous networks of cities, regions and private organisations have emerged to promote co-operation. Furthermore, several attempts have been made to organise, or reorganise, a structure more formal than that of inter-municipal co-operation (Larsen and Nousiainen, 1998). This process has been encouraged by the restriction imposed on regional policies by national governments, the decentralisation of new powers and responsibilities, and the threat of increased competition between cities and regions after the creation of the single market (The Commission of the European Communities, 1992). Competence is gradually being transferred from national states to regions and cities, and this has in itself increased differentiation in the spheres of public service provision and regulation of economic activity (Jensen-Butler 1996; Van Weesep, 1996; Van den Berg et al., 1993). As part of the shift in focus, the Principle of Subsidiarity is repeatedly invoked as the ideal to which such a process of transformation of tasks and responsibilities could aspire. The principle of subsidiarity (Rome Treaty in 1957 in Art. 235 and Art. 325 later the Maastricht Treaty Art. A (2), Art. B (2) and Art. 3b (2), Amsterdam Treaty Art. A (2)) represents a bottom-up approach, empowering the lowest level of competence in question. It has also been argued that the principle of subsidiarity represents a top-down approach where the set of tasks in question is specified by higher administrative levels. The point of origin for the principle of subsidiarity lies in the notion that action should take place at the appropriate level of Government, whether it be local, regional, national or supranational. One definition of subsidiarity can be found in Article 130 (r) of the Maastricht Treaty: "If and insofar as the objectives of the proposed action cannot be sufficiently achieved by the Member States and can therefore, by reason of the scale or effect of the proposed action, be better achieved by the Community" (Williams, 1996). According to this definition, the principle of subsidiarity concerns only the relationship between the Community and Member States, and not the relationship between society and state. It illustrates the fact that the lack of one overall definition the principle of subsidiarity leaves wide variety of interpretations, often to the detriment of overall goals. Higher levels are subsidiary, or ancillary, to the lower level (Gretschmann, 1991, Osborne and Gaebler,1993). Subsidiarity rests strictly on a dialectical relationship, where a smaller unit's right to act is operative to the extent that it alone is better able to act and achieve the aims being pursued than a larger unit. In this sense, lower levels of authorities and jurisdictions have precedence over higher ones. If a local level should fail to properly exercise authority, a higher level should not take over responsibility, but should instead help the lower one to cope with the problem on its own. *Pari passu* with decentralisation of responsibilities, centralisation of issues such as supranational importance takes place. Hence growing decentralisation of relevant tasks and responsibilities cannot take place without centralisation of tasks and responsibilities of a more complex nature. Accord between the principle of subsidiarity and, respectively, decentralisation and centralisation of responsibilities is unmistakable. The principle of subsidiarity can thus be seen as a double-edged sword which prevents both higher and lower levels from taking action in areas properly falling within each other's respective spheres of action (Schilling, 1995). The spatial extension of each task must in some form be quantified, however. The process is far from easy because it evokes conflicts, both horizontally (between municipalities) and vertically (between State, County and Municipal levels). These are conflicts about centralisation and decentralisation of powers and between vertical and horizontal administrative structures (political control, competence and power). Although difficult to manage it is important to determine at what level and which kinds of tasks should be located where.

The principle of subsidiarity provides no solution to the problems described above, nor can it be used as a guide to overcome scale problems. It presents only an a abstract goal rather than a method. It is a principle and not a rule. In every case, in every field, individual decisions must be made in order to achieve the proper level of governance. This may vary from region to region, or from city to city.

Solidarity on a spatial scale

The best way to make the welfare state function more efficiently is by decentralising responsibilities from higher to lower authorities (Dieleman and Mosterd, 1992). Furthermore, smaller units are often said to increase public participation and to be less expensive, even though the Dutch experience has shown the opposite development (Van der Veer, 1998). In theory, decentralisation to lower levels creates a higher degree of responsiveness, accountability and transparency for citizens and administrators, this is especially true if the system is geared to handle the changing circumstances, which is seldom the case. Furthermore decentralisation will lead to increased democracy, influence and local activity.

Decentralisation of responsibilities will proceed to the local and the provincial levels, as a logical extension of the reallocation of responsibilities. Van den Berg et al. (1993), contradicts with this view, argue that the scale of the local authority in which tasks are being decentralised, no longer corresponds to the social and spatial-economic problems it faces. Local municipal and provincial levels are unable to solve and exploit the opportunities or tackle the accumulated problems, either because of scale discrepancy or due to their lack of power and tools to do so. The result is a disproportion between problem solving and scale. Although decentralisation can fosters responsiveness, accountability, innovation, transparency and democratisation of the public sector, it is no guarantee that they will be achieved. Neither effectiveness nor economic efficiency in provision of public services will necessarily increase with decentralisation.

The principle of subsidiarity thus contains more than tangible considerations concerning scale versus tasks, it also entails an equally important abstract - solidarity. That subsidiarity consists of this dual notion is often overlooked but in fact it is the social aspect, where a functional urban area as a whole assumes difficult tasks, constitutes a component which is just as important as the more tangible economic argument. Basically, the issue is one of governance on the right scale with the right tasks (Engelstoft, 1994; Delors, 1991).

Subsidiarity should be applauded as a principle both in terms of economic performance and in terms of solidarity. If the attitude towards solving problems together (between horizontal and vertical administrative levels) and not individually is not successfully integrated, the result is not subsidiarity but merely a simple form of decentralisation. One could argue that from a municipal point of view, solidarity is good, while decentralisation of competence and power is better. This is often the attitude, unfortunately exemplified by NIMBY effects.

Simple decentralisation has the tendency to create polarisation, and polarisation creates despair among small units who, by the time the polarisation crystallises, will seldom have the power to alter their situation. The paradox of simple decentralisation is that responsibilities increase while the ability to solve structural problems decreases or the problems become even more difficult to solve. The dilemmas can only be resolved at a higher level of government, i.e., by a body with centralised authority and power. Once again the interconnectedness between centralisation and decentralisation is illustrated.

Solidarity, the desire to solve problems on a spatial scale exceeding administrative borders (both horizontal and vertical), creates the major difference between simple decentralisation and the principle of subsidiarity.

Subsidiarity deals with an economic approach towards administrative units and spatial solidarity. In a number of ways, the essence of the principle of subsidiarity makes it possible to combine the external and often global changes with internal democracy considerations. In general, existing structures are geared to operate under more orderly and intelligible conditions. It has been argued that FURs must pay greater attention to their competitive profile to meet and sustain a competitive position in a global or/and European setting. Furthermore, it has been argued that a necessary consequence of this requirement is a change in the scale and competence of administrative units. Most administrative borders and systems date back a long time. Changes in external and internal circumstances have placed new demands on the existing administrative levels. As mentioned, due to the lack of a unified definition, subsidiarity has often been used and abused for political purposes. We must guard against the potential for such abuse in some of the effects of using the principle of subsidiarity, which are of direct or indirect interest to politicians and civil servants (Larsen and Nousiainen, 1998).

The second wave of metropolitan government in the Greater Copenhagen Area

The Greater Copenhagen Area covers some 1.5 million people divided into 50 municipalities (fig. 11.1). Every municipality belongs to a county, except for the two central municipalities, Copenhagen and Frederiksberg, each of which performs the dual function of county as well as municipality. The three remaining counties make up the Greater Copenhagen Area (fig. 11.2).

Figure 11.1 and 11.2 Denmark and Greater Copenhagen area

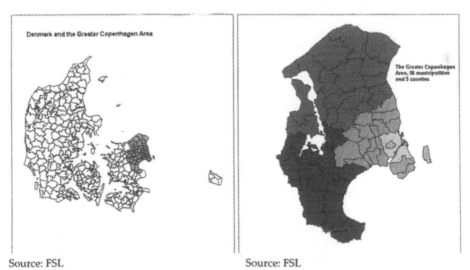

Source: FSL   Source: FSL

The division of jurisdictions creates major differences in required tax-based expenditures and income tax burdens in the various administrative units within the Greater Copenhagen Area. Local government expenditures are financed through taxes, user charges and state subsidies. Most county and municipal activities are financed by county and municipal taxes and user charges. The main principle is that the counties and municipalities finance expenditures if they can significantly influence the amount spent. Responsibility follows money and vice versa. The State pays 40-100% of the expenditures if the municipalities have little influence on the amount of the expenditure (Ministry of Environment and Energy, 1995).

Like other large metropolitan areas, the Greater Copenhagen Area has experienced several waves of regional interest for creating new, more comprehensive administrative structures in order to regulate and plan in a more efficient and effective manner. These different waves have not been especially successful due to resistance and the involved parties lack of willingness to acknowledge their shared responsibility i.e., (state, councils and municipalities). Much of the reluctance facing the previously elected governmental body (Hovedstadsrådet 1974-1989) originated with the additional administrative layers of government and not the citizens. A short presentation of the general administrative system follows. In Denmark, urban governance and planning are mainly a municipal responsibility. Although state and counties play a role in urban areas, it is the local municipality that is responsible for dealing with spatial development, planning, regulation and urban change. From a welfare and regulatory point of view, this responsibility places great pressure on the local level. Most decisions and discussions concerning the urban systems are taken at the local level, but due to scale and jurisdictional problems the best solutions seen from a overall point of view are seldom found due to the conflicts between welfare, efficiency and effectiveness. In some cases, it is not at all unrealistic to say that the municipalities are often in competition with one another, sometimes to the detriment of their citizens welfare.

The variety of tasks makes it difficult for citizens, organisations and other organs of administration to know who is responsible for what, creating the basis for poorly integrated policy-making. Many interactions between various departments of government and political parties overshadow the real relationship between government and citizens. The present administrative structure consists of a three-tier administrative system: state, counties and municipalities. In principle, each of the three levels of administration is free to conduct its own policy, provided that it does not contradict the policy or regulation of a higher authority. When municipal and local plans are made, they must be approved by the counties and the municipalities affected. Still, the division of authority is not as straightforward as it might appear. There are several hundred formal agreements for inter-municipal co-operation, as well as numerous formal bilateral agreements between administrations of different levels. Taken together, the agreements virtually guarantee the lack of a integrated policy for the Greater Copenhagen Area. Furthermore, the lack of transparency must be viewed as a democratic problem. The weakness of the Greater Copenhagen Area is that little or no comprehensive political agreement and mobilisation exists among the 50 municipalities and three counties. This is particularly true for an administratively decentralised country such as Denmark, where competence given to the municipalities enables independent policies. Because of the lack of a strong regional administrative body and the considerable power of the municipalities, development of a strategic comprehensive plan is virtually impossible. A few municipalities have benefited from the lack of an overall plan ("islands of hope in a sea of despair") but for the region as a whole, the lack of regional governance is far from an optimal solution. Both competitive parameters and social solidarity suffer from the lack of an overall plan. This has been acknowledge by the national government, the provincial authorities and the local municipalities in the area, at least on a formal basis. Several attempts have been made to reduce the negative externalities created by the lack of co-ordination.

In FURs such as the Greater Copenhagen Area, such conflicts become quite explicit because of the difficulties involved in setting up a cohesive and co-ordinated environmental and welfare-related effort. These conflicts are particularly unfortunate because the best outcome in terms of achievements and results would be gained from within large urban agglomerations. With the establishment of the Capital Commission (Hovedstadskommissionen) the national Government (represented by the Ministry of the Interior) proposed the reorganisation of the Greater Copenhagen Area in 1994. External developments played a major role in the government's argument. Globalisation of the economy, increased accessibility (technological as well as physical), and integration among the EU member states are mentioned as some of the reasons for creating a new division of power and competence within the region. Some problems and possibilities, it was argued, needed to be solved at a higher level of government. Others should be solved on a local level. According to the Ministry of Internal Affairs, the main proposal for a regional structure should ensure clarity and decision-making powers for all levels in question. The principle of subsidiarity was not explicitly mentioned in the proposed reorganisation, possibly because it is a very misunderstood and abused term. The position was expressed in an easily understood straightforward manner, but the arguments remained the same: to improve service provi-

sion, regulate spatial and economic activity and combat the negative externalities created in the region. In other words, the Capital
Commission would undertake responsibility for the regions future
development. The new administrative body would only deal with
strategic issues. Issues of non-strategic character would be dealt
with on a local level, "do locally what can be done locally and regionally what must be done regionally".[1]

The proposal presented three main characteristics for the new
structure: (1) strong political legitimacy, obtained by the direct
election of its political representatives; (2) meaningful autonomy
from both higher governmental and basic local authorities, acquired as a result of adequate financial and human resources; (3)
relevant territorial coverage, roughly consisting of the functional
urban region. As it often occurs in Danish decision-making processes two spatially different proposals were quickly produced: a
comprehensive model covering all 5 counties and 50 municipalities
(fig. 11.3) and a more modest model covering 3 counties and 21
municipalities (fig. 11.4). The competence and power given to the
new administrative body remained the same the only difference
was the spatial extension of the model.

Figure 11.3 and 11.4 Two proposals for Greater Copenhagen area

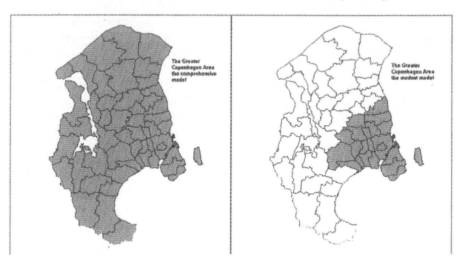

Source: FSL   Source: FSL

The politicians wanted to choose from two alternatives. Both
alternatives should give high priority to, and co-ordinate the effort

---

1   Motto from Stadsregio Rotterdam: *"Lokaal doen wat lokaal kan en regionaal wat
        regionaal moet"* (Larsen & Nousiainen, 1998).

on single assignments and on sector specific areas. Some special areas of interest were pointed out, e.g. spatial planning, urban development, health care and the placement and distribution of refugees. The regional body would be given supplementary powers rendering it able to overrule and decide over municipalities and counties whenever regional questions arose. From case to case, the regional body could decide what kinds of tasks should be located where, which in itself is not very democratic nor transparent in terms of the division of tasks and responsibilities.

Abolishing the regional proposal

The proposal for a new administrative solution for the Greater Copenhagen was rejected by the National Government (with broad support from Parliament) before any open or detailed discussion could take place. Although both the provincial and municipal levels had made several objections, these were probably not the reason for the Government's refused to implement or even conducting a debate concerning the new administrative body.

Even though it was emphasised by the Capital Commission in their report (1995) that a need for an integrated approach was needed in order to solve the problems experienced in the region, a broad majority in Parliament sought a more thorough examination of the division of tasks between the three administrative layers in general. Consequently the regional proposal was postponed, and instead replaced by a proposal for several different commissions who would each have to submit their recommendations by the end of 1998 and 2001, respectively. Two important commissions should be mentioned. First a Task Commission (*Opgavekommissionen*) of a complex political nature was established, with the task of evaluating the existing division of tasks and responsibilities between the three administrative layers, not only for the Greater Copenhagen Area but for Denmark as a whole. The Task Commission's work was clearly linked to the 1970 municipal reform (the major reorganisation in 1970, where 1388 smaller municipalities were reorganised into 275 new municipalities). Recommendations had to be submitted by the end of 1998, which occurred, but surprisingly without touching on the issue of administrative reorganisation. Second, the Government established the Capital Committee (*Hovedstadsudvalg*) which in 1996 consisted of eight ministers and their civil servants (in the meantime the number of ministers has been reduced to seven due to the fusion of two ministries). The main task of the Capital Committee is to monitor and evaluate developments in the Greater Copenhagen Area, mainly through organising and initiating discussions and analyses. The Capital Committee has no formal competence or power with regard to physical planning or regulation. Although the elimination of the proposal is still controversial, it is not an understatement to imply that national politics and strategic thinking formulated by the Central Government made any effort to continue with the proposed suggestion impossible. The process was initiated by the Central Government and was abolished by the Central Government, all due to "real political" considerations. We will not discuss this in detail, but merely state that in the process of policy formulation, declarations of statement, agreeing with general concepts for the benefit of all (e.g., increased welfare) are often met with general acceptance, because everyone can see and acknowledge the underlying rationality. The implementation of such strategies however, is a totally different story. Although there may be agreement and consensus towards general goals, and perhaps even agreement on the spatial policies required, it often proves difficult to implement such policies in urban regions, due to their many administrative levels, interests and political parties (Denmark's two largest political parties the Social Democrats and the Liberals had opposing interests in the process) and due to the need for co-operation between public and private sector bodies. There often exist conflicts of interest between the various administrative levels; for example, a desired development may mean that some municipalities are deprived of an opportunity to develop their business communities. As a result, central government policies, though acknowledged as sound from the point of view of the environment and sustainability may not receive very strong backing. Lastly, the example from Copenhagen has shown that purely national (Parliament level) politics may also play a role in influencing how reorganisation should take place within FURs.

In acknowledging the difficulties attached to the process described above, a new and very modest proposal has been suggested for the Greater Copenhagen Area. The Ministry of the Interior has proposed creating a Development Council (Udviklingsråd) for the Greater Copenhagen Area. The council will consist of eleven appointed members, of which five are permanent: the mayors of Copenhagen and Frederiksberg and the three council mayors. The remaining six members are be elected, two from the municipality of Copenhagen, two from the Copenhagen Council, and one from each of the two remaining councils of Frederiksborg and Roskilde. Focusing on solving the trans-municipal problems so important for the future development in the region, the new council will have to try and operate. The proposal cites the following key areas: cooperation in the Öresund region (interaction with south Sweden), regional physical planning, traffic regulation and planning, business development and tourism, and cultural development. The Development Council has jokingly been characterised as a "toothless mayor's club". Furthermore, by selecting members of the Development Council indirectly, the question of democracy and representation can once again be questioned, both in terms of the balance between the central municipalities vs. the smaller surrounding municipalities and in terms of giving a voice to representatives from the smaller political parties. Because participation can be conducted on a volunteer basis, it is quite clear that the new council is in no practical position to guide or direct the future development in any rational or desirable way - the instruments are simply not available. It must be stressed that the Development Council is still in the discussion phase - but it is likely to be passed through Parliament.

Barriers for successful alteration of existing administrative regimes

Several reports deal with the difficulties and barriers that administrative reorganisations can generate (cf. March and Olsen 1989). Although the explanatory basis of these analyses takes different approaches and uses different tools of explanations, they all deal with the transformation and conservatism of existing systems. Some authors seek their explanations in the formulations and actions of the principal actors involved, while others seek deeper explanatory complexes in the progress of understanding developments. In the following, a few points from the New Sociological Institutionalism will be presented.

Reorganisation of tasks and responsibilities in administrative systems has a number of implications for the existing structures. The primary problem is vertical transformation of competence and power either upwards (centralisation) or downwards (decentralisation), depending on the nature of the task. The essential point to emphasise here is that this shift of influence and power does not take place without resistance. Due to the rationality that lies behind the principle of subsidiarity, few will argue directly against it (here termed "formal politics"). That is why these kinds of reorganisations of administrative layers will often be meet with invisible or not fully articulated resistance, or "real politics" (cf. Flyvbjerg, 1992 or Larsen and Nousiainen, 1998). The rationales behind resistance are unclear or difficult to understand. Power, arguments, and tools to preserve them work in mysterious ways. Once the structures of power have been established and structurally embedded, it is very difficult to let go of them again. With power follows privilege and space for "manoeuvre" (both as an institution and as an individual). Which of the two forces plays the most significant role, institutional interests or personal interests? Often it is difficult to distinguish between the two. Whether institution or individual, it is quite natural to fight or struggle for more influence and power. It is in the nature of every organism. Consequently, when the power of an institution is limited or totally terminated, it not only influences the institution or organisation but also many people who make the institutions work. Even though some reorganisation processes are met with general acceptance and understanding, underlying interests play a more important role. Case studies from Denmark and the Netherlands show that these movements or merging of interests are not to be compared with formal arguments for or against, but are much stronger (Larsen and Nousiainen 1998). Van der Veer (1998), who recently completed a study of metropolitan governments in the Netherlands, concludes that resistance also depends on the age of the urban agglomeration. The historic metropolis (e.g., Amsterdam) is more capable of resistance than a new industrially based metropolitan area (e.g., Eindhoven).

' As some forms of resistance are impossible to see and thereby impossible to control or act against they often emerge when it is too late. The sum of real politics repeatedly proves to be too much for even the slightest alteration of existing regimes. Power defines rationality, not the other way around (Flyvbjerg, 1992). If these interests are not recognised and dealt with in a swift and orderly manner, they become uncontrollable. Overnight shocks, where systems are altered from one day to the next seem to be one way to overcome the articulation of interests. Yet even getting this far can prove to be more than difficult to control.

Concluding remarks

The main problem with the Principle of Subsidiarity's normative approach to reality is that it provides no potential suggestions as to eventual solutions to our problems. Consequently, the basic problems persists, both in terms of economy and solidarity. This is due to the fact that the Principle of Subsidiarity deals with formal politics and not the more intangible real politics. The principle thus locates itself together with other high ideals such as democracy, liberty, freedom. Unlike some of these ideals, however, the principle of Subsidiarity has yet not proven itself as a workable concept, although this chapter has shown that the Principle of Subsidiarity rests not only on the term rationality but to some degree on common sense. If we want to reorganise administrative structures, the Principle of Subsidiarity's second string lower priority to solidarity is perhaps the real Achilles heel. Why should smaller, more distant municipalities voluntarily take part in resolving the large deficit created by the central and often poor municipality? Higher levels of effectiveness and efficiency can be obtained if responsibilities are distributed to the "right" level. Decisions can then be taken on the level where the relevant information is found. It is widely recognised that when local units are empowered to solve their own problems, they function better than units that depend on services provided by outsiders. This is true for several reasons: the local level understands the problems better because this is where they are crystallised; smaller units are more flexible and creative than large administrative units. Lastly, it is often cheaper for society as a whole (Osborne and Gaebler, 1993).

What is often forgotten, however, is the level above the local level, the regional or state level. These levels must handle tasks of a more strategic nature. Here, too, empowerment should take place. What kind of tasks should be located where? The exact division of tasks and responsibilities between different levels of administration is not an easy task. How is it possible to determine the distribution of tasks between different levels of administration? A few points should be stressed here: for instance, that certain tasks are of a strategic nature, others clearly not; that a strong upper level of administration is just as important as an empowered local level; that one cannot exist without the other, as they are strategically intertwined and are partly a result of each other. Moreover, it should be remembered that subsidiarity ensures and maintains a diversified spatial structure, also with regard to social, cultural and political issues (Gretschmann, 1991). The exact division of tasks and responsibilities between local and regional levels must be decided in open discussion involving as many interests as possible (politicians, citizens, institutions and organisations). That this is easier said than done is obvious.

In the long run co-operation would also make it possible for municipalities to offer the best welfare for inhabitants within the boundaries of the region. The matrix of municipalities that together form a functional urban region must adapt themselves to the qualities of one municipality often being defined by the product offered by their next door municipality.

New regional administrative bodies within FURs offer, at least in theory, the possibility to combine both aims. By allowing the regional body to handle truly strategic regional matters, and local bodies to be responsible for truly local welfare ones, the possibility is open to deal with tasks at their proper level and on the right scale. This possibility must be understood not only in the context of subsidiarity but also in terms of purely economic benefits.

Metropolitan government in Europe

Nearly all large cities in Europe and in the world generally have experienced a growth outside the institutional definition of the original municipality. This raises the question of an instance of regulation at a pertinent geographical level. This question is not a new one: already in the 1970s, numerous books and articles were devoted to this problem. Two main models were utilised: to regroup some municipalities or to construct a new authority.

The merging of municipalities to constitute a metropolitan area has given rise to a number of objections, of diverse importance, which we can briefly summarize:

For some people, such a large unit will be too far from the preoccupations of the citizens, therefore causing some problems in fulfilling the traditional function of the municipalities, largely based on proximity.

The disequilibrium between a big center and the surrounding municipalities can be a problem. If there is a merger of municipalities with the center, this disequilibrium will be reinforced.

If urban sprawl continues, a merging will be only a temporary solution. In this case, it will be politically difficult to continually extend the boundary of the urban region year after year.

Such a movement was already observed at the beginning of the century. At this time, the center was rich and the surrounding municipalities poor. Nowadays, the small surrounding municipalities are often richer than the center and therefore have a strong interest in maintaining their privileges.

The creation of a new authority also has some drawbacks:

Very often, this will be a new unit, between region and municipalities. Such a new level is not necessarily easy to implement in a political system where those elected have developed careers and resources inside the existing institutions.

A metropolitan government is often designed by its component municipalities and thus escape the legitimating process of direct election. Furthermore, the question of political responsibility is not totally clear, and the capacity to modify the budget orientations is restricted for each of the component municipalities. Finally, the capacity of the citizens to identify themselves with such a delegated form of local government seems weak.

Analysis of these models reveals an interesting debate between a search for efficacy in administration and a quest for legitimacy by the local government. At the time, the balance between these aspects seems rather difficult to attain. Furthermore, these institutional rearrangements also imply a change in the distribution of power. In this sense, more attention should be accorded the different actors involved and their respective interests in order to have really successful experiences.

Dominique Joye

References

ANDERSEN, H. T. and JØRGENSEN, J. (1994) *Storbykonkurrence omkring Østersøen: København i nordeuropæisk perspektiv.* Arbejdsnotat 6. Geografisk Institut 1994, Denmark.

BERG, VAN DEN. L., VAN DER MEER, J. AND VAN KLINK, H. A. (1993) *Governing Metropolitan Regions.* Avebury: EURICUR.

CANADIAN URBAN INSTITUTE (1994) *Background paper with respect to the question on the Ballot City of Toronto Municipal Elections November 1994.* Toronto, Ontario.

COMMISSION OF THE EUROPEAN COMMUNITIES (1992) *Regional development studies. Urbanization and the functions of cities in the European Community.* Liverpool John Moores University. European Institute of Urban Affairs.

DELORS, J. (1991) The principle of subsidiarity: Contribution to the Debate. In *Subsidiarty: The Challenge of Change. Working Document proceedings of the Jacques Delors Colloquium 1991.* European Institute of Public Administration.

DIELEMAN, F. M. and MUSTERD, S. (1992) *The Randstad: A research and policy laboratory.* Kluwer Academic Publishers, Dordrecht.

ENGELSTOFT, S. (1994) *Regionalisering og subsidiaritet i Europa.* In J. Tonboe (Ed.). *Territorialitet. Rumlige, historiske og kulturelle perspektiver.* Odense Universitetsforlag.

FALUDI, A. and RUIJTER, P. (1985) No match for the present crises? Theoretical and institutional framework for Dutch planning, in, A. K Dutt and F. J. Costa (Eds.): *Public Planning in the Netherlands.* Oxford University Press.

FLYVBJERG, B. (1992) *Rationality and Power: Democracy in practice.* Chicago University. Chicago Press.

GRETSCHMANN, K. (1991) The Subsidiarity Principle: Who is to do What in an Integrated Europe? in *Subsidiarity: The Challenge of Change. Working Document proceedings of the Jacques Delors Colloquium 1991.* European Institute of Public Administration.

HARTOFT-NIELSEN, P. (1994) Ny stærk organisering af storbyregionerne i Holland - København kan lære af Rotterdam-modellen. *Byplan Copenhagen* 1994/5 Denmark.

JENSEN-BUTLER, C. (1996) Competition between cities, urban performance and the role of urban policy: a theoretical framework. In C. J. Jensen-Butler et al. *European cities in competition.* Avebury.

JØRGENSEN, I. (1996) Subsidiarity, Planning and the Concept of Region. *NORDRevy,* No. 5/6, 1996 pp. 23-26.

KOTLER, P., HAIDER, D. AND REIN, I. (1993) *Marketing Places. Attracting Investment, Industry, and Tourism to Cities, States and Nations.* The Free Press New York.

LARSEN, K.S. AND NOUSIAINEN, J. (1998) Governing functional urban regions - an inquiry into the needs for and difficulties in implementing new administrative structures - a Rotterdam experience. *Geographica Hafniensia,* Copenhagen Denmark.

LEFÈVRE, C. (1998) Metropolitan Government and Governance in Western Countries: A Critical Review. *International Journal of Urban and Regional Research.* Blackwell Publishers, Oxford UK. Vol. 22, No. 1. 1998.

MARCH, J. G. and OLSEN, J. P. (1989): *Rediscovering Institutions - The Organisational Basis of Politics.* The Free Press.

MINISTRY OF ENVIRONMENT AND ENERGY (1995) *The Urban Environment and Planning-Examples from Denmark.* Copenhagen. Ministry of Environment and Energy, Denmark.

NOUSIAINEN, J. and JØRGENSEN, G. (1997) *Bystruktur og Transportenergi: Metoder for empiriske studier.* Danish Forest and Landscape Research Institute, Hoersholm Denmark.

OSBORNE, D. and GAEBLER, T. (1993) *Reinventing Government. How the Entrepreneurial Spirit Is Transforming the Public Sector.* Plume England.

SCHILLING, T. (1995) Subsidiarity as a rule and a principle, or: taking subsidiarity seriously (revised version). Harvard Law School. Internet: http://www.law.harvard.edu/groups/jmpapers/schill/.

URRY, J. (1990) Conclusion: place and policies. in Harloe, M. Pickvance, C and Urry, J. (1990): *Places Policy and Politics. Do localities Matter?* London, Unwin Hyman Ltd.

VAN DER VEER J. (1998) Metropolitan government in Amsterdam and Eindhoven: a tale of two cities. *Environment and Planning,* Vol. 16. Great Britain.

VAN WEESEP, J. (1996) Urban policies to promote equity, in *European Cities in Competition* C. Jensen-Butler et al. (Eds.). Avebury 1996.

WILLIAMS, R. H. (1996) *European Union Spatial Policy and Planning.* Paul Chapman Publishing, Ltd.

12 Governing the City

JACQUES LÉVY

> "Toutes choses sont tuées deux fois :
> une fois dans la fonction et une fois dans le signe,
> une fois dans ce à quoi elles servent
> et une fois dans ce qu'elles continuent à désirer à travers nous"
>
> Julien Gracq, *Le rivage des Syrtes*.

This chapter will reflect on the possible conditions for a reform of territorial powers, which would create urban governments in a country, France, where such a project encounters a deep-rooted centralist tradition. France is used as an extreme example of the difficulties Europeans encounter when they attempt to equip their cities with relevant, transparent, and efficient governments. Like Switzerland and unlike the majority of other European states, France has barely changed the frontiers of its municipalities, which were carved out along the lines of the former parishes. Moreover, any reform in this area is extremely difficult due to the existence of a strong anti-urban stance in French political life and institutions, and the constitutional barriers erected a century ago. Thus, France differs from countries such as Germany, and Austria, where the federal structure encourages the emergence of a consistent, strong local power, and from those countries without explicit federal structures such as the UK, the Netherlands and Sweden, where strong urban cultures and traditions of reform through inter-institutional debate allow smooth changes.

The French case is therefore necessary as a *borderline* experiment. Indeed, in clear-cut situations, the results are not always satisfactory. This example can therefore clarify in a more general way, the problematic of urban government in Europe.

A legitimate reform

In France, the issue of the territorial reorganisation of powers has been cropping up, albeit *obliquely* in public debate. Sometimes the participants seem to be discussing something else: "municipal politics", "multiple elective mandates", sometimes the issue is dealt with from a technical angle: the "local tax system", "intercommunality", etc. It is clear, however, from the *Sueur* report (1998), the forthcoming consolidation of conurbation authorities (the *Voynet* and *Chevènement* laws, 1999), and from the increasingly insistent tune from DATAR, the French urban planning authority, that the debates are centred on a new association between political jurisdictions and geographical levels. This takes on special meaning at the city level for, since the 1982 decentralisation laws, communes have been playing a greater role in defining and handling urban problems.

This paper suggests that, whether addressed directly or indirectly, territorial reorganisation of powers is a major *political* issue. We shall analyse it from two distinct perspectives. First, we shall examine the logic behind a possible reform of the existing system. Second, we will look at the oppositions raised by the transformations under consideration. The two viewpoints allow us to characterise institutional change as both legitimate and improbable. We will conclude by comparing them in the light of the third stance of the research scientist delivering messages concerning politics.

In the debate concerning the reform of territorial division of powers, a distinction should be made between two moments: the moment when we ponder the need for one of these reforms, and the moment when we consider the obstacles such a reform would generate. It is appropriate in fact to examine the issue of the validity of creating a metropolitan power as such, without raising the difficulty of its implementation as a "prior issue".

Apprehending the city

The arguments in favour of a metropolitan government seem very solid. These arguments are based on a simple idea: every civilian society must have its political counterpart. Otherwise, decisions concerning the inhabitants of that society would be made by the government of *another* civilian society that might be bigger, smaller, or located elsewhere. Later on, we shall see that this is far from being a mere academic risk.

It is not easy to draw the demarcation lines of these local urban societies because, unlike states, which are separated by borders resulting from a balance of force between armies, cities are open spaces. While countries were busy consolidating the separating power of their borders (in Europe, between 1648 and 1945), cities were gradually losing their fortified walls and becoming mobile spaces, mingling territories and networks without clearly defined lines. Hence the existence of three types of difficulties.

First of all, urban areas are not easy to define. The notion of conurbation, which is administratively defined as a continuity of built structures, while quite close, lacks part of the urban reality towards the top and the bottom. Indeed, conurbations are not always contiguous and may include areas that are morphologically disconnected yet functionally linked by the labour market or commercial attraction.

This non-territorial but reticular (network) urbanisation is characteristic of the peri-urban phenomenon. Conversely, territorial continuity does not guarantee integration into a single whole, as is the case on the Mediterranean coast or in the Seine area downstream of the Paris metropolitan area.

Second, although the sizes of urban areas vary, they share certain essential features. There are both huge differences and fundamental similarities between a town of ten thousand inhabitants and a metropolis of ten million. The propelling force of the city is that, whatever its size, the concentration in a single place of maximum density and diversity offers vast groups of individuals, organisations and activities the possibility to meet and interact at any given time. Despite the swift development of transit systems and communications, which are the city's answers to the problem of vast distances, cities continue to increase their productivity gradient in relation to low-density areas. Hence, more than ever, the city exists as a concept and as a reality, and if we take a geographical and not geometrical or mathematical viewpoint, we are forced to admit that a small town such as Barcelonnette and a metropolis like Paris can be placed on the same *scale* in this case. Note moreover, that in terms of surface area, the differences are not as considerable as one would expect. Densities usually decrease parallel with the population decrease, which in fact means that small towns tend to spread out over large areas. Consequently, in two metropolitan areas like Paris and Bordeaux, the distances to be covered are roughly comparable although Paris is eleven or twelve times larger. In this context, cars play a larger role in small towns. If the statistics under consideration (time lost in traffic jams, pollution, etc.) raise the issue of transit systems in large cities, it is in small and mid-sized towns (of 50,000 to 200,000 inhabitants) that transportation raises serious problems of "exclusion from the urban life" for all those who for one reason or another do not have access to a car.

This does not mean that the same problems are raised in all cases, but that at least the basic question facing politics in these spaces - ensuring that the areas work as a daily context in which all factors are properly interlinked, without major conflict - is similar. It would be appropriate to fully accept this state of things and give each urban area the means to exist, irrespective of its volume. In concrete terms, this would entail defining the jurisdiction of urban areas, a definition that would apply to a small single-commune conurbation as well as to a metropolis. This jurisdiction, with its related tax resources, would necessarily include the search for major urban balances in housing, activities, and transportation. This would entail controlling housing policies, which would depend neither on the State (community housing) nor on the communes (building permits). All the urban development documents, from the zoning plan to the master plan, would be controlled by the urban council.

For Paris, it would be logical to define a conurbation that roughly corresponds to the boundaries of the current Ile-de-France region, and where the latter would be replaced by a Paris Basin region (Levy, 1998). However, it is quite conceivable that the Ile de France region would be given the jurisdiction of other urban councils.

A decisive aspect of this transfer would be the replacement of the Paris transport authority by a regional division of *Francilian* "government" transports. This division could decide to entrust the RATP, the Paris transit authority, with the administration of all the rail services of the metropolitan area including those currently run by the SNCF, the French railway authority. The RATP would then pay a fee to Réseau Ferré de France (RFF) and would integrate the entire large size network in an integrated metropolitan system, to the great relief of passengers who are tired and frustrated by the SNCF's lack of interest in its suburban lines.

Urban areas go through rapid changes. This is not surprising if one considers the dynamics of cities on the global scale. Cities are still very dynamic even in developed countries where the rural world reservoir has been almost completely drained. Recompositions within the urban world can significantly change demographic figures and relevant spaces within a few decades. This reality comes up against an important aspect of politics: territorial stability, which ensures that the citizens who decide and the inhabitants affected by these decisions are the same in the long term. It is not easy to administer a mobile space. There is the need for clearly-defined rules if one wants the rule of law to prevent "imperialistic" tendencies - a quite genuine risk as we shall observe below.

Defining the concept of "local"

Today, conurbations constitute *local* civilian societies, in other words, the lowest level at which the economy, social relations, geography and history interact to form a system. The need for "major balances" on this scale between habitat, activities and transport systems constitutes the mark and existence of a unified job market, a generally accepted indicator. Thus, the local character of a society is deduced neither from a surface area, nor from size of the population, but from the "everyday" character of its space.

This criterion is certainly not as simple as it might seem. The complexity and diversification of individuals' relations to time make it difficult to clearly define the "everyday", a concept that is increasingly fleeting. One of the solutions to the problem probably lies in the notion of *virtuality*. Fewer and fewer people do the same thing each and every day but, in a given context, we can strive to define the material potentialities and the ideal possibilities for action. The local area is then determined by all the social realities (individuals, material objects, and organisations) that are mutually accessible on the scale of the day, by all quasi-immediate encounters between all of its components.

Such a definition enables us to better distinguish between the different meanings of the word "local". Thus, the municipality within a conurbation, the neighbourhood (a notion requiring further explanation) and, more clearly still, the large complex or estate can in no way meet the definition given here of the local scale. Creating customs-free conurbations in the name of the principle that unemployment in a neighbourhood might be related to the absence of jobs in that particular neighbourhood, is an intellectually absurd and politically dangerous stand to take. It is common knowledge that the job market of a city does not work on this scale, and that the existence of jobs in a neighbourhood (for example a research centre in a housing estate inhabited by unskilled people) would have no effect on the unemployment rate. Indeed, this irrational micro-localism, creates illusions which undermine, in the long term, legitimate public action.

Similarly, the arrogance of some small town mayors who call their municipalities (fractions of a conurbation) "cities", is factually untenable and ethically unacceptable. A city is an urban society which accepts and organises its diversity to function as a dynamic system. But what do mayors do? Instead of organising "together-ness", they strengthen their respective vote-catching bases, and end up diminishing intermingling and increasing disparities. This is particularly striking in major conurbations, where dozens, some-times hundreds of municipalities destabilise major urban balances with egotistical public policies and undermine, by homogenising downwards or upwards, the perspective of social intermingling (Conseil regional d'Ile de France, 1998, pp. 38-39). Whatever the good intentions of these elected officials, the implacable logic of the administrators of these fragments of cities endowed with signifi-cant powers, is to "change the people" or strengthen pre-existing socio-economic trends. This dramatic experiment should enable us to understand the extent to which the decision-making scale is a decisive criterion.

An independent authority

Establishing a division of political cities based on a study of a more general division of societies is not an easy task. There is the risk of transforming much more complicated areas, combining territories and networks with hazy, mobile frontiers, into a country (a terri-tory marked by stable borders). However, it is difficult to imagine a political space that is discontinuous, incomplete, and volatile. In-deed, the political function requires the integration of all geo-graphical interstices whether or not they are used at a given time, and for a time period that extends to generations to come, in the-ory, forever. The territory is then clearly defined, and in an egali-tarian scheme, where all territories have the same rights, the politi-cal division then becomes *partition*.

The exercise thus suggested inevitably depends on the com-promise between contradictory requirements. This does not make it impossible nonetheless, and the contradictions themselves ex-press the real dynamics, and the partial asynchronies of a living society. Therein lies a question of political philosophy which is far more constitutional than decisional, and related less to the democ-ratic operation than to the very emergence of politics. City dwellers are to a certain point condemned to live together, since they share the same urban complex. By acknowledging this fact, we are sim-ply confirming the prevalence of the society's right over the *force* of existing authorities.

All experiments reveal that we cannot rely on the free decision of pre-existing political entities. Indeed, the richest or most power-ful will be the least motivated to join solidarity schemes with those who are poorer or weaker. For instance, giving a municipality the possibility to belong (or not) to a conurbation government, while it may seem democratic, is nothing of the sort. Rather it allows part of the urban society to secede and yet continue to take advantage of its position within the conurbation. This, then, would be anything but a democratic choice, since such a choice can occur only in an already defined society (where is the *demos* referred to?). Asking inhabitants if they are ready to live together with the other inhabitants with whom they already share a daily life is not a democratic exercise, less so if the elected representatives of political entities likely to lose power in the process are the ones asking the question. Politics is methodologically anterior to democracy. One can and must engage in politics democratically, yet remain conscious of all the consequences involved. The breaking away of parts of conurbations inhabited by people who wish to pay less tax, as sometimes occurs in the United States, (what I call the "Beverly Hills syndrome") would only be acceptable if the inhabitants of the "dissident" city agreed to stop being part of the conurbation (Lévy, 1994). Such a decision would be not only counter-productive, but practically impossible to implement.

In French conurbations, the division into small municipalities multiplies this phenomenon to several times in each conurbation, with spillovers effects on taxes (business taxes) as well as socio-political spillovers (stronger vote-catching bases and creation of wider gaps). The issues at stake can be expressed in terms of the following question: Should the former divisions (such as the parishes of the *Ancien Régime* or the *départements* created by the French revolution) have temporally unlimited power to impose themselves on the present? We would respond that the dictatorship of memory over a project cannot be considered as an established democratic principle.[1]

Under these conditions, who can propose and implement new divisions?

One can imagine INSEE and DATAR jointly developing a measuring tool that combines morphological and functional approaches, thereby improving the tools developed over the past twenty years.[2] Imagine this indicator being enacted; the frontiers of urban areas would be automatically adjusted at each census. Such a

---

[1]   This can be compared to the situation in the US, where the Republicans tried to make Congress approve the following: the voting of any tax increase would depend on the existence of a qualified majority, practically impossible to have. Former congresspersons – those who would have defined the previous budget revenue – would then have control over present and future congresspersons.

[2]   Particularly, Industrial or Urban Populating Zones and Urban areas. See the work of the VillEurope team on measuring urbanity and chapter 1 in this book.

procedure would prevent the negotiations, bargaining and exceptions that we observe each time an inter-communal space is created. Moreover, it would have the advantage of making urban authorities face their responsibilities. In other words, accepting a new urban spread by allowing an "uncontrolled" porosity between the urban area and its fringes would mean that the authorities had to accept new duties related to transit systems and infrastructure, without enjoying the transfers of resources from an old neighbourhood to another, since the game would be, at best, a zero-sum one. This might lead urban councils to monitor their zoning plans more closely.

One therefore wishes that the definition of the metropolitan space would be conducted as a report drawn up by a private agency created by law along the lines of independence of statistical institutes, ethics or even justice committees such as the CSM (*Conseil supérieur de la magistrature*), the French magistrates' council.

Democracy would then be practised on two levels. First, of course, through the direct election of the urban council, thereby ensuring genuine citizen participation on competing projects concerning the entire urban area.

Second, by law, voted at the political level immediately above that of target areas – here, considering the existence of transregional conurbations, the national level, in cooperation with the European Union for transnational conurbations – which would create the independent demarcation authorities, define their makeup and set their operating rules.

A perverse couple would then disappear. We are referring to the couple which in French political culture, associates a so-called uniformity (supposed to be a guarantor of equality) in the administration of the territory with an extremely differentiated practise so as to avoid the effects of general principles which are deemed undesirable. Contrary to today, neither a ceiling nor a floor would be placed on the size of cities. This is because, if we abide by the definitions proposed here, we notice cities of extremely disparate sizes in the "civilian" space.

An improbable reform

What are the obstacles to the implementation of a reform that would restore cities to their rightful place in politics? The first impression is that of an institutional inertia to change. Comparative analyses of existing urban governments show that, each time that the challenged pre-existing levels are powerful, their resistance is strong, even after the change has taken place and the trend can sometimes be reversed, as in Barcelona and Toronto (Lefèvre, 1993; 1996 and 1997). The rebirth of Greater London, after Margaret Thatcher eliminated it, also shows that nothing is final in the other direction. In Northern Europe, municipalities are generally more easily extensible than in the South, with France and Switzerland constituting borderline cases of the "frozen" communal mesh. It appears particularly clear that federal culture fosters territorial innovation, whether it is via the moving of communal frontiers or via the creation of new supracommunal bodies (such as the *Stuttgart Region*). The failure of the Berlin/Brandebourg merger, however, confirms the fragile nature of any change, which intrinsically becomes the focus of all oppositions.

Pushing the analyses even further, we notice that large-scale, protracted processes are at work in these resistances. The setting up of urban governments, indeed, raises fundamental questions to the territorial organisation of powers.

Immanence

Where there is disparity between civil society and political society, between those who decide (or give legitimacy to those who decide) and those who bear the effects of these decisions, we are no longer in the realm of *politeia* but in an empire. I have demonstrated, moreover, that these imperial situations did not concern a single but two main categories (Lévy, 1994). In the case of transcendence, one is in the presence of a traditional type of colony, the dominated territory constituting one of the pieces of the imperial entity.

In the case of immanence, on the contrary, one observes the dictatorship of the small over the big, in other words a sub-set of the political society imposing its decisions on the whole.

The latter is not as rare as one would think, particularly in Europe where the constitution of nations by States has been a mixture of military conquests, dynastic alliances and negotiations between rulers. In a system where each component of the national space would be worth its real demographic or economic weight, these regions would have been threatened by the risk of staying on the fringes for a prolonged period. Furthermore, the specific capacity of the inhabitants of the State's capital to influence national politics would have further widened the gap. To stave off the risk of immediate or future rejection of integration within it of regions with strong local identities, and increase the threat of a possible secession, States have had to be overly indulgent with these peripheral regions.

In Western Europe, three scenarios can be identified.

First, the situation where differences of development and influence were slight and the risk of frustration low; in this case, the periphery's influence on the centre has been limited. It is the situation of the Germanic world. In the German case, although all territories are represented in the senate (the current *Bundesrat*), the system for calculating representation is nonetheless only slightly different from the traditional citizen democracy of "one man, one vote".

The second type is more British but can be found in the Netherlands and Denmark as well. Forms of autonomy negotiated and given in response to specific demands allow the periphery to be contained within the apparently centralised but actually modest and open State.

Finally, in the more striking third configuration, the distribution of powers is profoundly affected by the weight of the periphery, which suddenly, acquires a paradoxical centralness. In Switzerland, the role given to the approval of cantons, irrespective of their population, on an equal footing with the "people", for any major decisions, ensures a decisive control for the Germanic mountain regions with their "primitive" cantons, which admittedly, have never been peripheral from a political viewpoint. In Sweden, Norway, and Finland, the sparsely populated and barely developed Northern regions enjoy considerable tax and state policy advantages. In Italy, it is the entire *Mezzogiorno* that seems to be a disproportionate player, still involved in a client-based political system where voters expect to received advantages in exchange (*scambio*) for each vote cast.

In its own way, France combines all the features of this third type. Immanence exists on the municipal (untouchable frontiers, up to the absurdity of municipalities without inhabitants), departmental (division into cantons giving a differentiated weight to "urban" and "rural" voters in county council decisions) and national (senate and the single ballot elections per district for the legislature) scales, with a general tendency for national political choices to drift towards the local. Lastly, not only Corsica, but also all "State" cities and regions enjoy considerable financial transfers (Daveizies, 1997).

1-3/2-4

How does all the above translate into urban governments? In the third type, infra-urban scale political spaces are often equipped with, on the one hand, considerable powers, and on the other hand, constitutional barriers to prevent the system from changing. And they are rarely willing to abandon either.

A simple principle can be used to classify territorial players: 1-3, 2-4. As demonstrated by François Moriconi-Ebrard in his comparison of the urban networks of several countries, the more centralised a political system (level 1), the more it tends to reinforce, sometimes by creating from scratch, a level 3, made up of small, frail entities (Moriconi-Ebrard). This is the case of the communes+departments system in France. On the contrary, the level 2, that of large cities (including Paris), is maintained in a maximum state of political impotence. This configuration summarises fairly well the history of the distribution of powers in France for the last 100 years. The founding compromise of the Third Republic sets up this system and uses institutional keys and the senatorial bolt referred to above to lock it up.

The emergence of new legitimate political scales, Europe, and more latently, the World, changes the deal. If the level 1 is no longer national, possibilities of unusual alliances have cropped up and make the results less immediately foreseeable. Just like the regionalists, partisans of strong urban governments quickly understood the support they could find in the European way of thinking, which is encountering the same obstacle on its route: resistance to state control.

This being said, the struggle remains unequal. As we can see with the resistance of elected officers to the restriction of the number of elective mandates, only sufficient pressure from public opinion would be able to break down the institutional barriers.

Governance: the politics-free democracy?

Before we can reflect on its democratic operation, urban government raises a problem of the constitution of a political space, a *politeia,*. It is clear that democracy does not lie in the constitution of the political society but in its administration.

To say that government must precede governance is not to preach a "return to the State" as some would wish. Others, who fear it, have so very well internalised the state's refusal to share sovereignty that the only way they can imagine a strong infranational power is as something surreptitious and interstitial, as if "politics with a capital P" were in advance and forever, impossible at the levels where armies and foreign policy do not exist. This is a confusion between "sovereignty" and "majesty", according to Leibniz's distinction. Cities may have a genuinely legitimate and democratic government. By this term, we mean that the government *stricto sensu* is but an element of a larger process which does not reduce citizens to mere voters, which combines the informal and the institutional, participative democracy with representative democracy and counter powers with powers. Faced with the crises of the political link, which has hit vulnerable social groups, a great deal of imagination is required to avoid amputating one of the limbs of the political society. The professionalisation of political life inherently contains a distance from civil society, and any attempt to prevent this confinement by involving players defined by their economic or sociological function more directly in political life would be welcome.

All this is well and good, and should be applied at all levels, and neither more nor less to the city than to the others.

In particular, this list shows by itself the need, as a pillar and as a reference, for a formal component of governance, the government. Failing this, the idea of responsibility, that of governments towards the governed, that of citizens in the face of their own choices – would disappear, dragging down in its wake the very idea of politics. The paradox would be asking for more governance from the urban area even while it lacks government, to attempt to heap flesh onto a boneless body.

We can therefore not content ourselves with praising the "flexibility" of the current multiple inter-communal configurations. It is a good idea for communes (or districts of large metropolised municipalities) to play the role of counter-power, in other words, to check the powers of infra-local bodies carrying the opinions of part of the urban society, faced with a deliberative mechanism (legislative and executive) operating at the level of the conurbation. However, such a function is antinomic with that which consists of proposing and implementing a method in order for a society to manage its diversity in a productive manner. An institution cannot work as both an interest group and as a *polis*, a political society. With its crossed financing and undermined decision-makers, excessive "flexibility" can end up muddling the stakes and messages, preventing the transparency of the society, a primary condition of any political life.

Legitimacy, improbability and action

What is happening? Things may be changing but the transformations are few and slow.[3] Can we hope to jump the precipice in dif-

---

[3]   In his article "Le territoire, une idée neuve en France", *Le Monde*, 13 November 1998, Jean-Paul Besset believed that the convergence of forces and the

ferent stages? Perhaps. There are still a few obstacles, some of which may be bypassed and some ignored. We must nevertheless observe that although it may seem to be on another level, the emergence of a powerful "environmental awareness" modifies the *factors of production* of a legitimate debate on space. This has brought about a new consensus that is, on the whole, favourable to a better acknowledgement of urban reality. There is still the risk that a new institutional stratum, superimposed on the existing one (without changing anything) and hastily devised (the local powers are left to battle over the reforms) will result in the establishment of a useful perspective and an increase in the general incoherence of the system.

Let us sum up the situation. A reform consisting in creating a veritable government is perfectly justified, not only because it seems reasonable from the outside, but also because it conforms to the principles established by the political society: state of law, democracy and justice. Why, then, does a legitimate reform remain improbable?

Who is afraid of urban democracy?

Some people still continue to fear urban democracy for reasons of their own.

They are those who have built their electoral legitimacy around the reinforcement, in fractions of a city, of a monolithic sociology. Fear of urban democracy also applies to those who, in this age of a subsidiary Europe and an interdependent world, still dream of an all-powerful government and who, to preserve this myth, will do anything to avoid the emergence of strong players, as would inevitably be the case with new urban councils.

Lastly, it is the case for those who preside over county councils and distribute the money of urban dwellers without asking for their views to create, in the name of a nostalgic "ruralness" and by unscrupulously holding several elective mandates and functions, an electorate that is strictly a vote-catching one. In France, rural societies have disappeared, and there exists no local society still functionally structured by agricultural production and its consequences in all sectors. There is, however, a "political rurality": a movement which defends he advantages enjoyed, within the political system, by the beneficiaries of the privilege of *territory* (i.e., the surface area, even if it is unpopulated) over the *democratic principle*. Given the institutional apparatus for sharing territorial pow-

---

concomitance of projects can effectively set the reorganisation of territorial powers into motion.

ers, this "party" is not insignificant because it is reinforced by peri- and infra-urbanisation. The inhabitants of sparsely populated areas can effectively judge how much it is in their interest to defend the power of municipalities, the *département*, the district election and the Senate, since they can observe that this automatically goes to fill the coffers of local authorities under whose jurisdiction they live.

There are political movements and social groups in France, therefore, which in the name of ideological convictions or sectional interests, view any questioning of the *single scale* unfavourably. This scale is that of the nation-state that associates an idiosyncratic domestic state control with a prominent geopolitical attitude outside its borders. It is not surprising that this generates a resistance to sharing *governmentality* between different levels ranging from the local to the global.

In Europe, however, there is also a strong current of eschatological thinking. By this, I mean the attitude consisting of disclosing and denouncing the evil forces that threaten society to a point where they may provoke an inevitable disaster that can only be diverted by a revolutionary uprising. This stance is quite a noble one. It is a result of both the tragic dimension of Prometheus' gesture – the revolt against the Gods is futile – and its Judeo-Christian extension – only the Last Judgement will bring about a revival of mankind.

By offering the possibility of an approach that is reactive without being, at least in principle, reactionary, this attitude provided European *progressivism* with a considerable part of its resources, over the ages.

If we apply this attitude to our problem, it is oriented in three directions that are seemingly contradictory but are in reality complementary.

1. Institutional changes are impossible;
2. Institutional changes are of no use;
3. Institutional changes are dangerous.

The first proposition is based on the acknowledgement of the power of constitutional barriers mentioned earlier on in this article. If we add the conviction that the most determined key players will mobilise to fight against any reforms, then the cause is resolved. Only a revolution can set a situation that has come to such a desperate standstill into motion again. The second proposition tells us that the real problems are to be found elsewhere, in a participatory or direct citizenry, in business or even more fundamentally, in unemployment and exclusion. It also tells us that focusing on "formal" issues diverts public debate from more critical questions. The third proposition aims at defending the pivots of society's most threatened sectors (suburbs, rural areas) in the present political configuration as a result of the power of municipalities and *départements*. These sectors would become even more disadvantaged once the declaration of the democratic principle results in making the weak even weaker.

The rejection of politics

What is really common to these three assertions is the rejection of politics. The Rousseauean faction of the Enlightenment chooses natural law over justice, justice over democracy and democracy over politics. For politics to exist, there must be citizens who are capable, even if only during the time of a public debate, of freeing themselves from their own personal conditions to involve themselves in a desired common future and to find the methods required to attain this. Progressivism of this type is necessarily gradual. It does not believe in a sudden and final reversal because it rejects the idea that the responsibility for the present situation can be attributed only to a conspiracy by those in power and that this conspiracy can merely be identified and then eliminated.

It is not so simple to think of oneself as a citizen among other citizens. Since 1789, French society has existed in an ambiguous relationship with politics. Rural societies, through their local particularities, followed by the working class and other social groups, considered the national political scene to be a distant place which had to be pressured if one's complaints were to be heard. The flow of legitimacy between the society and its representatives, we must admit, is a very abstract system from which the common citizen cannot expect to attain immediate and direct results. The political bond remains just as unfulfilled in vote-catching systems or in the lobbying by interest groups and communities as seen in the United States. In essence, politics, when seen from the point of view of a human being or a group, is always presented as a diverted path where the advantages obtained by each person are only a by-product of the overall advantage to be obtained by society. This relationship with politics is not compatible with community allegiances, which stand in the way of the direct insertion of the individual into political society. The more society has actors who are dynamic, aware and independent, the less credible becomes the attitude which takes responsibility away from individuals and blames blind structures.

The seemingly technical nature of the debate on institutional reforms masks a fundamental controversy in which defenders of the prerogatives of the centralised Government and the self-styled representatives of the "dominated population" who are keen on keeping their electorate, concur in rejecting the construction of a truly political dimension to social life.

Creating the actors, thinking like an actor

From the viewpoint of those who act either as researchers or experts, the effects of this controversy are quite significant. Forecasting is always a tricky exercise. One of its complications is the difficulty in assessing the effect of the forecasting message on the people at which it is targeted and consequently, the modification in their behaviour on the phenomenon under study. In the case of the "Unacceptable scenario" published by the DATAR in 1968, the experts were explicitly counting on the reactions of political players to prevent the fulfilment of their sinister tale (De Roo, 1988, pp. 55-56). This hope for a negative reaction assumes the existence of powerful operators, who can change the course of events but are intellectually limited. To convince them, reality must be caricatured, and they must be excluded from the script. They are thus made to react to a play in which they play no part (Lussault, 1993). This example, that we today observe in the typically French approach of the European Spatial Development Perspective (ESDP) that was launched in the European Union during the Liège summit in 1993, corresponds to an intermediate plan of action between two extremes (Levy, 1997/1998, chap. 8). At one end, nothing is expected of the players and they are informed only of what is going to happen to them. At the other end, they are presented with a situation full of contradictions and thus with different degrees of freedom of action.

We must ensure that we are in the right "play". If we underestimate or overestimate the capacities of the actors, there will inevitably be a mistake in our forecasts. This may lead, through mere caution, to a distrust of "scenarios". Here, the ethics of expertise is closely linked to the search for efficiency, which means contributing to making an event real by merely announcing that it is inevitable, i.e., a "self-fulfilling prophecy".[4] In such a scenario, the freedom of action of the other actors is saturated by means of the overwhelming legitimacy of one's own margin as an expert.

It is always risky to disarm players by telling them that whatever they do, the result is predetermined. This is even more so when these actors, whether individual or collective, do exist and

---

[4]   See on this point, concerning urban dynamics, the thoughts of Jean-François Staszak.

are ready to take over the problem. Of course, the opposite mistake, consisting in foreseeing a social uprising, which turns out to be pure fantasy, is also serious. Beyond the necessary prudence, we must not surreptitiously transform improbability into necessity.

Many major events in history, particularly the most spectacular, would have been considered highly improbable by any forecaster.

Sherlock Holmes' famous method ("When you have eliminated the impossible, whatever remains, however *improbable*, must be the truth") takes on new meaning in our contemporary societies. This is because we encounter different types of forces and mobilisations capable of achieving the "improbable": they provide an unexpected, rapid and powerful link between the civil and political sides of the social world. This is especially the case in France, which lacks a good bottom-up flow of legitimacy through the official representation channels, leading to seemingly spontaneous movements (i.e., among students) who end up teaching the players on the political scene a thing or two. In a society of actors, not everything is possible, yet we can advance the hypothesis that the distortion between cognitive tools and political debate is less inevitable or at least less durable. In the cases discussed here, the political legitimacy of the project stems from the meeting of a purely intellectual rationality (it is logical to compare the civil society and the political society) and an ethical approach (this comparison is good for the society). The latter may never be reduced to the former, simply because they are not of the same nature and do not convey the same messages or reality. However, we can consider that the *translation* from one register to another can be made easier in a context where the difference between the professionals of knowledge and *ordinary people* has shrunk, in terms of knowledge of the social universe. Short of claiming an essence that is profoundly distinct from that of their fellow citizens, no scientist may proclaim that the path leading from the intellectually necessary to the politically legitimate is definitively blocked.

This is why we can conclude that the legitimate transformation of the French political landscape is neither probable nor impossible. It depends on the quality of another space that is metaphorical but decisive: the "public space". Here we fond not only the pivot, but also the stakes. Urban governments are expected to create an appropriate framework for the implementation of public policies. They are expected, in particular, to provide a forum where citizens meet to debate the kind of urban model that they want realised (Levy, 1997).

Urban sprawl, the gradual domination of the car, the zoning of functions, the loss of intermingling, the threat to old centres and the springing up of "emergent centralities" on the periphery of urban centres have all been simply endured. In other words, they have acted on the *micro* scale of causes without considering the *macro* scale of effects. On the same level as this tendency remains on the sidelines (otherwise called the "Johannesburg model"), there is a resurgence in the market for urban ideas, of an "Amsterdam model" that prioritises density, diversity and complexity. The second model seems to be more an unstable political configuration than the result of a clear and shared choice. At present, the reversal of trends in favour of a "European" model of large urban balances is based mainly on the emergence of an environmental awareness. Likewise, most of the public policy tools that hamper periurbanisation were primarily devised for the protection of the environment and not as a result of a public debate on the desired urban model.[5] The first task of an urban government is not to solve problems, however, but to ask its citizens questions. One such question could be: "Is what this the kind of town that you the actors wanted to have?" Or: "if there were several urban models to choose from at the same time and at the same place, how could we organise their cohabitation in such a manner that tomorrow, you, the citizens of this town, would still be able to decide which one of them is the most unacceptable to you?"

---

[5]   Below are the main environmentally friendly laws, with their effects on the urban landscape. Mountain law (1985) and Coastal law (1986): urbanisation control; law concerning Water (1992): encouragement to reduce the extension of built-up areas; Landscape law (1993): limitation of unplanned constructions; Environment protection law (1995): regulation for city entrance points; law on Air (1996): Plans for moving within the urban area.

The city against the region?

Throughout this book, and especially this chapter, we have seen how the geographical scales concerned with urban problems constitute a complex but important topic. We need to identify the pertinent scales of the urban phenomenon as well as changes. From a morphological point of view, we have already spoken of urban sprawl and Citta diffusa throughout this book, but we have also stressed that urban changes are related to economic and social processes, primarily globalisation.

Actually, to be acknowledged as a world city the main cities perceive themselves in competition and no more situated in a complex center-periphery net of relations. This could be seen in the very interesting work of Cattan et al., when the authors try to measure the actual factors of centrality. This is also a process that influences the creation of urban networks, trying to insert common resources to attain a critical influence. In the same sense, urban marketing is a means to valorize the resources of the cities, not in the sense of endogenous development, but to place them on a global market. Some infrastructures have a significant role in this context, primarily international airports that define a new geography, based on their accessibility. However the transition to a knowledge-based economy is also a factor favouring the centrality of the main urban regions. Some researchers, Klaus Kunzmann for example, propose to considering some urban regions of about 100km radius, corresponding to places that we can access within the limit of one hour.

Such a new geography represents not only a change in the economic or social life of the regions but also a main political challenge. The traditional hierarchy of local, regional and national levels will be totally modified if urban regions represent a new power, with more economic resources than the regional one, but reflecting another definition and another equilibrium: urban regions represent economic strength and, sometimes, social problems. Non-urban regions seem poorer when the traditional regions form a more regular pattern. The challenge of some cities, in having their problems taken directly into account by the national government or even by the Commission in Brussels without going through the official hierarchy of power, is a sign in this direction. If, as we have seen in this chapter, it is difficult to put into practice a new authority for the agglomerations, it will be even more difficult to solve the question of the political regulation of these new regions.

Dominique Joye

References

Conseil regional d'Ile-de-France, L'Ile-de-France. Réalités présentes, questions d'avenir, Paris, IAURIF, 1998.

DAVEZIES L. (1997) "La cohésion fragmentée", *Pouvoirs Locaux,* No. 33.

DE ROO Priscilla (dir.) (1988) *Atlas de l'aménagement du territoire,* Paris, DATAR/La documentation Française.

LEFEVRE C. (1993) Analyse comparative des institutions d'agglomération dans les pays industrialises, DATAR, prospective group "Territoires et institutions".

LEFEVRE C. (1996) "De l'intercommunalité donctionnelle à la surpacommunalité citoyenne" in Institut de la Décentralisation, *La décentralisation en France,* Paris, La Découverte, 1996.

LEFEVRE C. (1997) « L'Europe et ses villes : de la recherche du bon gouvernement métropolitain » in Serge WACHTER (dir.), *Des villes architectes. Retrouver les voies de l'urbanité,* La tour d'Aigues, L'Aube.

LEVY J. (1994) *L'espace légitime, Sur la dimension spatiale de la fonction politique,* Paris, Presses de Sciences Po.

LEVY J. (1997) "Penser la ville : un impératif sous toutes les latitudes", *Cahiers d'études sur la Méditerranée orientale et le monde turco-iranien,* No. 24, July-December, pp. 25-38.

LEVY J. (1997/1998) *Europe : une géographie,* Paris, Hachette.

LEVY J. (1998) "Quelle urbanite pour la région de Paris", *TIGER* (Travaux de l'Institut de géographie de Reims).

MORICONI-EBRARD F. (1993) *L'urbanisation du monde depuis 1950,* Paris, Anthropos.

13 Epilogue: Globalization and the City

ROBERT A. BEAUREGARD

Globalization looms large in contemporary urban theory. The rapid diffusion of telecommunications, roaming of transnational corporations, proliferation of multi-national trade arrangements, and increased international migration might or might not be at the root of current transformations in cities and urban systems. Nevertheless, urban theorists and researchers find the notion of globalization to be quite useful. From this initial position, they move easily to the novel dynamics and conditions that characterize the urban world of the late 20th and early 21st centuries. Each move, though, brings with it the tensions that exist between global and local forces; that is, between the institutions that transcend political and cultural boundaries and the particularities of place.

Not all of the authors in this book begin explicitly by paying homage to globalization, but all evoke it. The dynamic of European integration is palpable as each turns his or her attention to urban form, public space, and issues of governance. Each of the themes becomes meaningful at the overlapping spatial scales of European realignment, and each takes on even greater salience in the stark contrast between globalization and localization (Joye, Jorgensen and Ostendorf, this volume).

The purpose of this concluding essay is to place the preceding chapters in the context of globalization and reflect on the convergences of urban form, public space, and governance within western Europe and between the cities of Europe and those of the United States. Globalization brings with it fewer boundaries to the spread of ideas and the movement of investors and households. Yet, it also engenders a reaction to the accompanying homogenization. Specific cities attempt to distinguish themselves from the global crowd and social groups resist a lost of identity and history. Engaged in competition with cities in their region or across the globe, cities invest in buildings and events that enable them to be compared favourably with their competitors even as they quest after a unique international image. This tension is one of the defining traits of the contemporary city.

Urban form

Urban form is linked to globalization through the greater emphasis on the competitiveness of city-regions, through the threat that globalization poses to local identities and the historical distinctiveness of individual cities, and through the diffusion of architectural styles and real estate projects that comprise the symbolic vocabulary of global success. Accompanying the globalization of the past few decades has been a neo-liberal ideology that subordinates the national state to global capitalism, discourages extensive welfare provisions, and encourages place competition. One result has been a severing of national financial supports for local governments. This has forced localities to be more competitive in order to attract investors and households and to raise tax revenues. At the same time, people are threatened by global forces that seem indifferent if not antagonistic to the cultural particularities embedded in specific places.

City governments have responded by searching for ways to compete in the global arena and yet maintain the city's distinct history. One path they have taken is to attend more closely to the quality of their built and natural environments. Cities have worked with investors to construct new urban places -- festival marketplaces, historic districts, museums, entertainment complexes, among others -- to generate a sense of vibrancy and heighten their attractiveness. Signature buildings, such as the Guggenheim Museum in Bilbao, epitomize this confluence of global competitiveness and the worlds of international real estate and high culture. Cultural policy, tourism, financial services and technology are the bases on which cities compete for investment and image (Andersson, this volume; Cantell, 1999).

For many cities, the goal is also to enhance the desirability of the city as a place to live. Parks, safe neighborhoods, waterfronts developed for residences and recreation, and well-designed public spaces are part of a strategy that emphasizes environmental sustainability and the quality of life in the city. In a world where elites and educated labor can locate wherever they wish, many argue that they will be attracted to cities where urbanity reigns and amenities are abundant (see Mitchell, 1999).

The emphasis on sustainability and quality of life has been made possible, in part, by a transformed global division of labor that relocated heavy industry outside the advanced economies of western Europe and the United States. In their place has emerged activities (for example, business and financial services, health care, education) that require educated workers, pay well, and are less environmentally damaging than traditional goods-production. Low-wage jobs still exist and many workers are still engaged in manual labor, but manufacturing no longer drives urban economies in the cities of these countries. A high quality of life and a "clean" environment are more desirable and more attainable in such settings.

It is not just the inner city that is being transformed by new global forces. Clearly, the inner city (or central city in U.S. parlance) is not alone an agent of economic growth. The city-region is "the" urban economy and only strong and integrated city-regions will fare well in international competition. Yet, urban regions are themselves in flux.

We are witnessing the emergence of new metropolitan arrangements. The concentrated dispersion labeled "citta diffusa" (Matteus and Governa, this volume) is a striking departure from the compact city (Ellefsen, this volume) that had previously characterized many European urban places. It compares closely to the multi-nucleated city-regions of North America, Los Angeles being the prime example (Soja, 1983; Dear and Flusty, 1999). All differ from previous urban patterns in western Europe and the United States. In the former, cities were relatively compact and spatially distinct. In the latter, the postwar norm was high-density central cities surrounded by low-density, mainly residential suburbs that created the infamous suburban sprawl.

Although American-style suburbanization has most often been viewed as environmentally destructive and economically and fiscally inefficient, the multi-nucleated alternative does not seem to be a noteworthy advance toward a sustainable city. The compact city, though, has its problems and attempting to re-capture the "ideal" through densification (Skovbro, this volume) is not assured of success. Cities are environmental disruptions (Rogers, 1997) and it is difficult to minimize and impossible to eliminate all environmental damage.

These different regional models have profound implications for daily life. The geographical juxtaposition of homes, workplaces, shopping areas, and places for recreation literally dictate the travel patterns of residents (Acquist, this volume). Women, specifically, bear the brunt of low-density, diffused urban form, as they bear most of the responsibility for child care and the management of the home. At the same time, women are participating more and more in paid labor and this exacerbates the logistic problems posed by low-density urbanization. The amount of time that women (and residents generally) spend in automobiles has significant environmental consequences; automobile emissions are one of the major contributors to lowered air quality and highways one of the biggest users of land (Kay, 1997).

Public space

Lurking beneath the issue of urban form and the concern with settlement patterns is the meaning of the urban. We can hardly comment on what has changed in the urban system if we cannot distinguish it from what is not urban (Joye, Jorgensen, and Ostendorf, this volume). Yet, this basic distinction eludes us as each conceptual assault (whether it be by density, function or culture) withers in the face of criticism and as the physical types of urban areas proliferate and meld together, continuum-like. Our intellectual formulations are also plagued by the meanings attached to the city by those who experience it on a daily basis. Objective and expressive meanings intersect as they are filtered through everyday language (Sayer, 1984). Consequently, urban researchers often fall back on political territory and statistical definitions devised by governmental census bureaus in order to delineate the urban, even though each harbors unavoidable distortions (Martinotti, 1999).

If any space is considered quintessentially urban, though, it is public space (Lofland, 1998). In European countries as well as the United States, to say public space is to refer directly, even if implicitly, to the city. Cities are comprised of workplaces, neighborhoods, shopping districts and an array of public spaces: parks, plazas, boulevards, monumental squares, sports stadia, and riverfront walkways as well as streets and playgrounds. Here, people come together, crowds form, and strangers co-mingle. The public space is one of the city's dominant images.

Public space, though, is not simply local. The re-discovery of places where people can congregate becomes more important as people become more mobile, uniformity spreads across cultures, and local political institutions turn increasingly outward in order to maintain economic competitiveness. The fascination with public spaces is thereby neatly joined to globally-driven, economic growth. In public spaces, tourists linger; alongside them, global businesspeople dine and make deals. These are the spaces of postcards and magazine advertisements. They are the raw material of city marketing. Here, strangers contribute to the local economy. Such spaces attract visitors and portray an image of urbanity, another resource in the city's quest for global recognition and stature.

That these spaces are not solely for tourism and publicity -- that residents also use them -- often makes them contested spaces. Consequently, theorists have become interested in how people live their daily lives in the face of global forces that undermine it and how residents of the city can become active in controlling what happens to them (Jorgensen, this volume). The latter leads to issues of participation and governance. In the face of large institutions that think and act transnationally; local institutions, groups, and associations become more central to our lives, as does the democracy that knits them together.

Public spaces thus take on a political dimension (Beauregard, 1999). Democracy requires that people of diverse backgrounds come together: march in solidarity, protest, witness a cultural spectacle, or simply serve as an audience for a political speech. People must be able to mingle freely and deliberate freely with others and in public.

Even the public spaces that are theoretically accessible to all are not used by everyone (Joyce and Compagnon, this volume). Filtered by who we are, perceptions mediate our use of plazas, parks, and playgrounds. Consequently, many public spaces become less than catholic; they become parochial spaces. The city is thereby arranged into distinct "social" areas with public spaces reflecting the social composition of the private spaces that surround them. Democratic theorists worry that such public spaces will become too "protected" or even privatized as people cluster only with those like themselves. By prohibiting or discouraging strangers from entering them, we narrow our understanding of others and of the city. In addition, we do psychological damage. Interaction with strangers requires that we empathize and compels us to reflect on who we are and how we live. Not doing so stifles maturity (Sennett, 1970).

Diverse social interaction is served by greater accessibility. Spatial arrangements are important here. The form of certain cities lends itself to accessibility while that of others encourages and encodes segregation (Levy, this volume). Johannesburg is not Amsterdam, and the low-density sprawl that is Phoenix (USA) compares unfavorably with the compactness of Oslo. American-style suburbanization and the citta diffusa seem inhospitable to copresence and interaction. The public quality of the city depends in

part on integrated transportation networks, geographical compactness, and the minimization of spatial barriers.

This concern is most pronounced in the United States where
mostly affluent and white suburbs contrast sharply with central
cities in which the poor and racial minorities are concentrated. The
isolation of these groups, one from another, is seen as undermining
a sense of awareness and empathy of the advantaged for the disadvantaged. Urban commentators call for regional solutions: metropolitan-wide subsidized housing programs, regional tax base
sharing, and metropolitan-wide land use controls, for example
(Rusk, 1999). The idea is to integrate the currently balkanized city-
region in order to enhance the public sphere.

Depending on the age of the city, many public spaces are the
spontaneous products of local cultures, serendipitous rather than
planned, and have long histories and entrenched and layered identities. During the late 19th and early 20th centuries, they became
the responsibility of local governments. In the late 20th century, we
are witnessing the emergence of "less public" spaces that fall under
the control of private-public partnerships, such as business improvement districts (BIDs), or private corporations such as the
management companies of shopping malls. One cannot think seriously about public spaces without accounting for the institutions
that maintain and manage them.

Governance

For this reason, any understanding of the urban form and public
realm of contemporary cities must pay heed, as the authors here
do, to public institutions and the management and planning of
cities. In one sense, this move is ironic. Globalization is widely assumed to weaken local institutions, making them more dependent
on transnational elites, multinational corporations, and a highly
competitive global economy. Global financial institutions such as
the World Bank and International Monetary Fund have been ag-
gressive in pushing a neoliberalism that supports this new dependence.

At the same time, the ostensible rise of city-regions as the
nodes of global commodity chains has elevated local and regional
political institutions (Parkinson, 1991). As part of being competitive, local and regional governments have to be stronger; they have
to negotiate with multinational corporations, produce an efficient
and technologically-advanced city, and deliver a high quality of
life. Urban management and planning remain important, particu-

larly in countries (such as Sweden and The Netherlands) where national involvement in spatial development has always been strong.

Urban management and planning link back to issues of urban form and public space, the environment, and participation and governance. State planning has consequences for everyday life and for the urban growth and sprawl (Ostendorf, this volume) that threatens accessibility and adjacency. The sustainable city, moreover, is hardly attainable in the absence of state intervention; sustainability and an unregulated capitalism seem irrevocably incompatible. The sustainable city is equally beyond our reach without the commitment of local citizens to collective goals and collective action, as in Local Agenda 21 (Basten and Lotscher, this volume).

Thus, managers and planners are expected to consult with the people affected by their plans. Planning means planning <u>with</u> the people not simply planning for them. The principle of subsidiarity is part of the dual concern with locality and democracy on the ground (Nousiainen, this volume). Decisions and implementation must be made to seek their "lowest" level in the institutional hierarchy.

Yet, one wonders about the efficacy of planning in cities populated by global actors and faced with the porosity of national, regional, and local boundaries. Managers and planners can neither wholly control the "space of flows" (Castells, 1989) in which their city is situated nor amass the resources and political power that enable them to realize their plans. We see this clearly in Ostendorf's (this volume) case study of new town and compact city policy in The Netherlands.

Issues of governance, public space, and urban form, of course, existed prior to our recent recognition of globalization's current manifestation and will exist after it (and our understanding of it) changes. Nevertheless, globalization places these issues in a different context, posing novel practical considerations and new interpretive problems. The meanings of the city are not the same now as they were fifty years ago. The city has changed as well.

Convergence

Lingering within the discussion of globalization and its local correlates is the question of convergence. Is globalization making cities and urban systems more similar from one country to the next? Are national differences being dulled?

Certainly, the dissolution of trade borders, the cessation of passport controls, and the adoption of a common currency speak to a greater volume of flows of commodities, capital, and people within the European community. In part, simply thinking that a European urban system exists is a sign of convergence, at least intellectually, though this image can hardly be considered unique to the late 20th century (Hohenberg and Lees, 1985). Convergence is also being purposively created, as witnessed by various policy initiatives (Albrechts, 1997; Kunzmann, 1996; Marlow, 1992). Much of this, of course, applies mainly within western Europe. As regards western and eastern Europe, the fall of the Berlin Wall in 1989 and the break-up of the Soviet Union serve as symbols of the spread of capitalism into once-hostile realms and the opening of borders. All of this suggests a convergence of ways of life as well as the dynamics of development.

Not all barriers are down, of course. For example, real estate development companies generally confine themselves to a single country or region. In doing so, they draw on national modes of financing, building technology, and social requirements. The built environment is still mainly made up of buildings and infrastructure designed for local activities and thus highly susceptible to local cultural and political influences. To the extent that real estate is de-localized and de-commodified (Beauregard and Haila, 1997) and international corporations penetrate local property markets, this generalization has to be qualified. Nevertheless, urban form changes slowly because it can only be changed incrementally.

My limited knowledge of the European countries keeps me from going further. The chapters in this book do point to an intellectual convergence, one encouraged by the European Union. The interaction of urban scholars from various European countries has produced agreement on common themes and a sharing of literature. At the same time, scholars are still entranced mainly by what happens within their countries and read mainly in their own languages. Dematteis and Governa (this volume) imply that the citta diffusa is unique to Italy, while simultaneously pointing out that many other countries have similar "labels" (for example, hypercity, ciudad dispersa) if not similar phenomena.

Notably, comparative analyses are less common than within-country case studies. This suggests that scholars still view their national urban conditions as sufficiently unique to justify their full attention. Regardless, this enduring differentiation indicates that convergence might not be as strong a force, at least as regards intellectual production and the urban development that it represents, as the forces of globalization and European regionalization imply.

I am more confident about my ability to comment on the convergence issue as seen from the perspective of the United States. Western European countries seem to be adopting a model of urbanism that has characterized the U. S. after World War II: auto-dependent suburbanization. Moreover, many large European cities have given rise to what also had been seen as mainly U.S. phenomena: edge cities and the resultant multi-nodal metropolitan areas (Lehrer, 1994; Prigge, 1998). Still, the dominance of social housing, the existence of planning mechanisms that encourage higher densities than in the U.S., and a tradition of apartment dwelling are forces that resist any full imitation of the U.S. model. Notwithstanding, many of the novel transformations of contemporary cities can be found in the United States, western Europe and Asia (Beauregard and Haila, 1997; Haila, 1999).

European local governments also seem to be following the U.S. model as regards local economic development. More concerned with competitiveness and, in some cases, forced to consider how to bolster local revenues, they have developed policies to attract high technology industries, corporate offices, and research activities. Quality of life, tourism, and cultural policy are also part of the economic development agenda. Large-scale, real estate projects (for example, Potsdamer Platz in Berlin) seem more common and public-private partnerships are on the rise (Parkinson, 1991).

Many European cities engage these initiatives and related private sector actors in the context of a strong national state, a much more politicized environmental movement, and still-significant welfare provisions. Property tax competition and government fragmentation, moreover, are not as serious problems as they are in the United States. In addition, although ethnic tensions (particularly around immigration) are not absent, the enduring racial divide that burdens the U.S. has been avoided. Social polarization and exclusion are still more pronounced in U.S. cities than in western European cities (Body-Gendrot, 1996).

The quest to be globally competitive has not stifled participation, although participation is often tightly controlled when issues of economic growth are being addressed with large-scale investments -- at least in the United States. In both western European countries and the United States, participation in urban management and planning is quite common. This was not the case in the U.S. forty years ago. Participation, then, consisted solely of consultation with local elites. Today, "western" planning models are, for the most part, participatory models, with corporatism more pronounced outside of the United States.

Finally, urbanism is more of a public issue across the globe than it was twenty years ago. Driven by prosperity and heightened affluence in the U.S., attention is being given (once again) to sprawl and problems of accessibility. The interest in urban form and sustainable development is also a reaction to the sterile forms and elitist procedures of high modernism, the rapaciousness of unfettered development, and the uniformity and homogeneity that globalization threatens.

In western Europe, the debate centers on the sustainable city versus the compact city. In the United States, the issue is the New Urbanism with its emphasis on town centers, mixed use districts, less reliance on the automobile, neighborhood socializing, and nostalgia for small town living (Lehrer and Milgrom, 1996). New Urbanism, however, is mainly a reaction to low-density and auto-dependent, postwar residential suburbs and offers only weak guidance for ameliorating the entrenched problems of the American inner city. Its European counterpart is the neo-traditionalism to which Prince Charles gave his imprimatur. Lacking traditional suburbs, western European countries seem disinterested in the New Urbanism. The global flow of ideas (and architects and planners) might be producing a convergence in architectural styles; urbanism, though, cannot be made fashionable so easily.

At the same time, many differences remain between the cities and urban systems of the United States and western Europe. The national state in the U.S. has always been a weak state, localism is much more entrenched, and our system of intergovernmental financing weakens territorial redistribution. Any transformation of the U.S. urban system, moreover, must contend not only with a vast country and an elaborate system of cities, but also a commitment to suburbanization that will be difficult to overcome. For the last two decades, U.S. cities have enjoyed a resurgence and their image has become correspondingly much more positive. City life is more competitive than ever with suburban life. Nevertheless, the culture of the United States has always embodied a deep ambivalence about cities (Beauregard, 1993). The city is simply not central to the national identity.

Conclusion

One of the many strengths of this collection of research papers is its ability to support the raising of fundamental questions about globalization and cities and about the global convergence of urban transformations. This is a book anchored in a strong and rich con-

ceptual framework in which urban form, public space and governance are intertwined. None of these themes, though, can be fully
understood independently of contemporary processes of globalization. Globalization changes the local; it also reconfigures intellectual space. The way that we think about cities has been transformed (Body-Gendrot and Beauregard, 1999).

262 Change and Stability in Urban Europe

References

ALBRECHTS, Louis. (1997) "Genesis of a Western European Spatial Policy?" *Journal of Planning Education and Research*, 17, 2, pp. 158-167.
BEAUREGARD, Robert A. (1999) "Julkinen kaupunki" ("The Public City"), *Janus* 7, 3, pp. 214-223.
BEAUREGARD, Robert A. (1993), *Voices of Decline: The Postwar Fate of U.S. Cities*. Oxford: Basil Blackwell.
BEAUREGARD, Robert A. and Sophie BODY-GENDROT. Eds. (1999) *The Urban Moment: Cosmopolitan Essays on the Late 20th Century City*. London: Sage Publications.
BEAUREGARD, Robert A. and Anne Haila. (1997) "The Unavoidable Incompleteness of the City," *American Behavioral Scientist* 41, 3, pp. 327-341.
BODY-GENDROT, Sophie and Robert A. BEAUREGARD. (1999) "Imagined Cities, Engaged Citizens," pp. 3-22 in Beauregard and Body-Gendrot (1999).
BODY-GENDROT, Sophie (1996) "Paris: A 'soft' Global City," New Community 22, pp. 595-606.
CANTELL, Timo. (1999) *Helsinki and a Vision of Place*. Helsinki: City of Helsinki Urban Facts.
CASTELLS, Manuel. (1989) *The Informational City*. Oxford: Basil Blackwell.
DEAR, Michael and Steven FLUSTY. (1998) "Postmodern Urbanism," *Annals of the Association of American Geographers*, 88, 1, pp. 50-72.
HAILA, Anne. (1999) "The Singapore and the Hong Kong Property Markets: Lessons for the West from Successful Global Cities," *European Planning Studies*, 7, 2, pp. 175-187.
HOHENBERG, Paul M. and Lynn Hollen LEES. (1985) *The Making of Urban Europe: 1000-1950*. Cambridge, MA: Harvard University Press.
KAY, Jane Holtz. (1997), *Asphalt Nation*. New York: Crown Publishers.
KUNZMANN, Klaus. (1996) "Euro-megalopolis or Themepark Europe?" *International Planning Studies*, 1, pp. 143-163.
LEHRER, Ute Angelika. (1994) "Images of the Periphery: The Architecture of Flex-Space in Switzerland," *Environment and Planning D*, 12, pp. 187-205.
LEHRER, Ute Angelika and Richard MILGROM. (1996), "New (Sub)Urbanism: Countersprawl or Repackaging the Product," *Capitalism Nature Socialism*, 7, 2, pp. 1-16.
LOFLAND, Lyn. (1998) *The Public Realm: Exploring the City's Quintessential Social Territory*. New York: Aldine de Gruyter.
MARLOW, David. (1992) "Eurocities: From Urban Networks to a European Urban Policy," *Ekistics*, 59, 353, pp. 28-32.
MARTINOTTI, Guido. (1999) "A City for Whom? Transients and Public Life in the Second-Generation Metropolis," pp. 155-184 in BEAUREGARD and BODY-GENDROT (1999).
MITCHELL, William J. (1999) *e-topia*. Cambridge, MA: The MIT Press.
PARKINSON, Michael. (1991) "The Rise of the Entrepreneurial European City, " *Ekistics*, 59, 351, pp. 299-307.
PRIGGE, Walter. Ed. (1998) *Peripherie Ist Uberall*. Frankfurt Campus Verlag.
ROGERS, Richard. (1997) *Cities for a Small Planet*. New York: Westview Press.
RUSK, David. (1999) *Inside Game/Outside Game*. Washington, DC: Brookings Institution Press.
SAYER, Andrew. (1984) "Defining the Urban," *Geojournal* 9, 3, pp. 279-285.
SENNETT, Richard. (1970), *The Uses of Disorder*. New York: Vintage.
SOJA, Edward, Rebecca MORALES, and Goetz WOLFF. (1983) "Urban Restructuring: An Analysis of Social and Spatial Change in Los Angeles," *Economic Geography*, 59, 2, pp. 195-236.

Printed and bound by CPI Group (UK) Ltd, Croydon, CR0 4YY

22/10/2024

01777627-0010